Atlantropa and Beyond

Magical Macro-Imagineering Mega-Projects

Richard Brook Cathcart

Infinity Publishing

ISBN 978-1-4958-2018-2 Paperback

Published July 2017

INFINITY PUBLISHING
1094 New DeHaven Street, Suite 100
West Conshohocken, PA 19428-2713
Toll-free (877) BUY BOOK
Local Phone (610) 941-9999
Fax (610) 941-9959
Info@buybooksontheweb.com
www.buybooksontheweb.com

TABLE OF CONTENTS

PREFACE

Homo sapiens, us, are the Earth-bioshell's least timid covey of like-minded creatures. Some humans are courageously bold, sometimes recklessly ill-equipped with our signature organized and interdependent technologies, boldly trying to ponder expansive geographical ideas, and to accomplish heroic deeds. During a Geologic Time period that is 1/30,000th the estimated duration of our known Universe, human beings have achieved a variously-assessed "ceno-Anthropocene" Earth Geologic Time period, at least according to many modern scientific and eco-social cliques. Some technologically advanced cultural groups, possibly representing all members of our learning and imagining species, have projected reporting machines into interstellar space as well as physically manipulated individual atoms! Certain sub-groups are attempting to create life-forms smarter than us and even life-forms that are to be first synthesized by us!

ATLANTROPA AND BEYOND: Magical Macro-Imagineering Mega-Projects is a book intended as a round-up of many ultra-gigantic, even startling, postulated macro-engineering mega-projects indicating, possibly, how our 21st Century and beyond probably might unfold in terms of global and regional human civilization's widespread, multi-planet infrastructures. Derived from the Ancient Greek-language word "*macros*" to indicate a thing is extra-ordinarily large or unusually long; therefore, "macro" implies significance, complexity and challenge; unfortunately, it also invites geographical and spatiographical quantification and therein is the essential terminological difficulty in formulating a perfectly clear and understandable definition! "Imagineering" is an U.S.A.-originated word, currently mostly associated with Walter Elias Disney, famed public entertainment film and comic-book creator. So, the combination of "macro" with "imagineering" results in "Macro-Imagineering", most bi-elementally means "BIG" thinking by human individuals, or concerted groups of individuals, about future changes intended to be made to the real-world which is constantly increasing in spatial volume. In inverted terminology, the real-world, in this instance, is a surrounding bubble—ranging from the known Universe down to the smallest technically created enclosure!

Our known Universe may have begun its presence with a "Big Bang", a name coined circa 1949 by world-famous astronomer Fred Hoyle (1915-2001) (Kragh, 2012) as a label for the largest conceivable explosion, nowadays represented, in part, by its 1965-detected cosmic micro-wave background (Bonnet-Bidaud, 2016). When we recall images of plants or roots pushing through concrete, it reminds us that life is energy! Whenever the "species" Homo sapiens perfects its working environment—today, only the Earth's so-called bioshell [incorrectly, its biosphere]—our species may flourish, increasing in numbers, health and wealth. Perfection requires careful maintenance of our species' immediate and extended material and energy environment, in the best applicable possible meaning of that admittedly very broad term. ["Species" is a Latin word meaning "form" or "appearance", and there are currently at least 30 different bio-scientific working definitions for that much-bandied but quite vague term (Wilkins, 2011; Hill 2017); such also is the indefinability case with "life", whatever that may be! And, what is truly "synthetic life" (Sauro, 2017)?] Popular Earthly bioshell use of the ceno-Anthropocene notion, wherein humans are classed as a degradative force of telluric amplitude, also demands an illusion, the biotic fixity of what is a true specimen of Homo sapiens: in other words, if humans soon live on Mars, change biologically through adaptation, some persons even becoming cyborgs or co-share equally an industrially developed Red Planet with AI robots, should Mars's "Anthropocene" be better named? A New Martian civilization won't be powered by fossil fuel extractions! Indeed, will there even be a discernable "anthros" concept helpful in defining the new Geologic Time period imposed upon Mars by Homo sapiens' arrival during mid- to late-21st Century? Would a spin-off from Richard R. Landers' ***Man's Place in the Dybosphere*** (1966) result in a Mars "Dyboshell"? If humans and AI robots achieve ultra-sociality [inter-"species" social cooperation] and shared intentionality, that really is a New World biotic-robotic potential physical action outlook! Some astronomers countenance the "day" eventually to come when humans and/or their non-biological "offspring" might migrate from our fully known Universe, through some kind of veil-barrier [the increasingly detectable details of the innermost "surface" of the cosmic micro-wave background?] into a parallel Universe—in essence, overcoming the confining power [energy, matter, time and whatever else] of our Universe's edging "eggshell", our region of space-time, which protects its contents with yet-undefined strength beyond merely mechanical! Famously, Si-Fi writer Arthur Clarke (1917-2008) presciently averred: "Any sufficiently advanced technology is indistinguishable from magic".

Macro-imagineers, generally speaking, have inclusive desires: for humor, Art and all forms and formats of tested certain knowledge and data, as well as their most significant signature human and humane expressed psychological characteristic: audacious macro-imagination skills. The amazing power of their imaginations, both individually and collectively, is possible because, as humans, they have been given God's gift of spirit: if it can be dreamt and perceived, then it can be daringly achieved! Every macro-problem has an equitable solution, and every human goal contains a means to achieve it; humans must, perpetually, prepare for positive results gained by thought and labor. So, please, dear reader, enjoy these popularly documented and worded, referenced literature chapters, carefully honed by Team Geographos being, we hope fervently, stimulation to project your own peaceful worldly desires and secret celestial thoughts onto our ever-increasing-in-volume universal and, possibly, post-Universe worldly bubble! Attempt a zealous foray!

REFERENCES CITED

Bonnet-Bidaud, J-M. (2016) The Diffuse Light of the Universe: On the Microwave background before and after its discovery: open questions. *Foundation of Physics,* DOI: 10.1007/s10701-016-0056-1.

Hill, G.E. (2017) The mitonuclear compatibility species concept. *The Auk: Ornithological Advances* 134: 393-409.

Kragh, H. (November 2012) Naming the Big Bang. *Historical Studies in the Natural Sciences* 44: 3-36.

Sauro, H. (9 March 2017) Enter the living machine. *Nature* 543: 178.

Wilkins, J.S. (2011) Philosophically-speaking, how many species concepts are there? *Zootaxa* 2765: 58-60.

Dedication

This book is dedicated to the pleasant memory of Edward William Thomas (1 September 1949-28 January 2017), professional magician, Past National President of The Society of American Magicians [2002-2003], and co-worker-friend.

Rest in Heavenly Peace.

CHAPTER ONE

Before We Ever Navigate the Universal Cosmic Micro-Wave Background Veil

Anthropogenic pneumatic structures are emblematic of continuous change in our Earth-world and the known Universe surrounding humanity (Steiner, 2005). The rapidity of everyday news and Science media-reported changes, worldwide and regional, stimulates our world's populace to appreciate public and private conservatories as vital bulwarks in the ongoing human effort to preserve plant and animal biodiversity. Some geoscientist claim Earth's bioshell is at a global Geologic Time vitality tipping point, the ceno-Anthropocene, and that this all-too-commonly apparent planetary life change is evidenced by authoritative rejection of an artwork's creation without any obvious official misoneism. Often it has been bruited that it is impossible to insult an elephant, because those creatures are so large in volume and quite thick-skinned; in fact, elephants are sometimes called pachyderms, from the ancient Greek word "pachys", meaning thick, and "derma", meaning skin. Nearly a decade ago, on 15 March 2008, in Santa Ana, State of California, the renowned soap-bubble artist Fan Yang intended to surround a bulky packyderm, the then 39 year-old 4,013 kilogram live female Asian elephant (Elephas maximus) affectionately named Tai, within a gigantic soapy orb for 5-10 seconds at the Discovery Science Center's annual Bubblefest Event. The controversial visual stunt was cancelled after vociferous local public complaint against the soap-bubble Science (Cohen et al., 2017) entertainment's unscientific, and possibly cruel, planned public demonstration! Previously, without any perceptible negative public outcry whatsoever, Fan Yang had gently placed 8 to 30 adult humans, on two entertaining occasions, inside singularly capacious soap bubbles in Santa Ana on 7 April 2004 and 12 April 2006. Such odd urbanite 2008 acquiescence to pervasive politically correctional doctrine is possibly indicative of ever-changing, sometimes contradictory, and even chaotic radical Green-inspired jumbled economic, political and sociological factual realities of modern-day Americans; in other words, with the final effect of generating an almost cartoonish, blankly empty real-world "free-speech balloon"! During 2016 Ringling Bros and Barnum & Bailey Circus stopped their popular American tradition of living performing elephants, substituting puppets instead! Ersatz elephants! [The Czech bubble-artist Matej Kodes, during January of 2017, surrounded, but did not fully enclose, 275 healthy young European persons—none of whom were sick persons, such as those who have died hermetically confined by plastic bubble materials, suffering from aplastic anemia (Ted DeVita, 1962-1980; DeZern et al., 2017) or severe combined immune deficiency syndrome (David Vetter, 1971-1984)—within a 11 meter by 7.5 meter rectangular floor area vertical soap screen for a few seconds. Actual physical bubbles are minimal surfaces assuming the characteristic globular shape with the least possible surface area to contain a volume of gases, inert and living objects. Soap bubbles are toys, possibly best created outdoors.] Asian elephants, on a mass-specific basis, metabolize at rates similar to bacteria and human beings (Gillings and Paulen, 2014). Humanity's first Earth-circling satellite, **Sputnik I**, a Russian word meaning "Fellow Traveler", was a shiny sphere launched on 4 October 1957! A sealed air bubble was the center-piece of novelist T.C. Boyle's Space Age-set ***The Terranauts*** (2017); the first space-flight of Yuri Gagarin (1934-1968) aboard a portable short-term life-maintenance chamber, **Vostok 1**, on 12 April 1961, made him an internationally-venerated spationaut. **Figure 1**, below. For some philosophical artists, the intentionality of a 43,000 kilogram mechanical elephant, "The Sultan's Elephant", promenaded through the streets of London, England as processional Art during May 2006 after its assembly in the hometown of Jules Verne (1828-1905), is the true meaning of 21st Century Earthly life, with robotic creatures possibly as Homo sapiens' inevitable intellectual off-spring and operational replacement (Sheil, 2008; Fukumori, 2017). Team Geographos, however, takes an additional step forward, we apprehend a human-inhabited spaceship or undersea vehicle's airlock as a threshold marking a discontinuity between working spaces (Jutte, 2015). **Figure 2**. Celebrated for their general as well as their specific task-memorizing abilities, friendly elephants are truly symbolic of Earthly life's universal macro-problem of retaining and communicating accurate information about our inhabited world and the known Universe which totally envelopes it. Those persons who consider the obligations of life to be "a vale of tears", "a dream", "a joke", "a game" or a "computer simulation programmed by vile extra-terrestrial Alien zoo-keepers" (Afshordi et al., 2017) may not see the true, essential adorableness of elephants quite as more thoughtful persons do.

Figure 1. Sculpted into a single boulder, the face of spationaut Yuri Gagarin, framed by his faceplate-less pressure-suit's helmet, sits on the southern shoreline of Lake Issyk-Kul in Barskooon, the northeastern section of Kyrgyzstan, and commemorates his safe landing after humankind's first successful low-Earth orbit mission. Symbolically, it represents all so far found life-forms emerging from the Earth-crust during our planet's Geologic Time. (Google Image.)

Figure 2. Boeing and NanoRacks LLC intend to construct the first privately-funded commercial air-lock for the International Space Station so that multiple deployments of small satellites can be easily accomplished during a single air-lock cycle from AD 2019. (Google Image.)

Earth Nature-originated, or randomly dropped onto or imported from elsewhere, pre-ceno Anthropocene life had its own seafloor Internet, and still does (Burdorf et al., 2017); humanity's Internet, buried beneath the seafloor mimics Nature's (Durairajan et al., 2015)! Yet, an electro-mechanically simple bubble-shaped—more accurately, globular or nearly so—artificial outer space satellite, the USA's imitative **Vanguard I**, launched successfully on 17 March 1958, is the oldest anthropogenic planet-orbiting object (Millard, 2017). The USD$ 100+ billion price-tagged International Space Station (Field et al., 2014), with its center of orbital plane coinciding with the center of Earth's mass, is a "station", like ground-based radio and television broadcasting facilities, that can unify and network Homo sapiens' nearly 200 hundred ecosystem-nations because it is a globalized participatory process-event, which people unequipped with instruments of any special kind can attend and enjoy visually—by the simple physical act of looking upwards, toward outer space's star-twinkling darkness—or by tuning into designated radio and television broadcast frequencies. Nowadays, non-functioning anthropogenic space debris surrounding Earth's bioshell might be characterized as a shell-shaped "bubble" of unwanted, sometimes despised [radioactive or poisonous], stuff that must be destroyed in a costly future Earth-orbit zone space clearance operation, most easily by air-friction induced incineration. The risks to life associated with unnatural space debris is increasing and forward-thinking Chinese macro-engineers aim to mop the orbiting non-conglomerated mess by use of an engine that converts space debris into a propellant, enabling the machine to maneuver almost indefinitely as it destroys man-made junk (Lan et al., 2015)! Until that might be accomplished, or some other effective means is utilized, the Earth's ceno-Anthropocene bioshell interminably endures a process-event of intentional international littering and globalized air pollution [deliberate satellite de-orbiting] rather like the futuristic intentional global and regional climate control magical mega-projects proposed by Geo-engineering's vastly multiplying public advocates (Low, 2016); R.B. Cathcart (2013) has continuously and vigorously advocated in favor of the regular bombardment of Earth's upper-atmosphere by the comminuted remains—minute olivine dust particles—taken from olivine-rich asteroids (Borisov et al., 2017), extracted by daring humans and drones to safely effect a planned globalized drawdown of strictly human-defined "excess" aerial carbon dioxide gas content that might cause some undesirable climate regime changes, and that as an operational mega-project does not pose any worldwide threat whatsoever to persons or real-estate properties! Since 1945 humans have created radioactive fallout and deliberately released various chemicals into near-Earth outer space from rockets and, as well, created the anthropogenic space hazard of space debris—the man-made waste, junk, and defunct objects orbiting nearby (Reasoner, 1992). Bombardment events might even become tourism's outdoor shows, just like the predictably periodic Leonid and Perseid meteoroid streams present to viewers!

Currently, living and breathing human spacefarers must anticipate an increasing spacecraft as well as human spacesuit integrity threat posed by ever-accumulating, and fracturing by collisions, anthropic space debris. Some spationauts rocketed or aerospace-planed into close outer space may not always be literary geniuses, but all of them must be vigilantly litter-wary (Ocampo and Klaus, 2016; Koppelmans et al., 2017)! The aggregation, shattering, and volumetric expansion (spreading above the turbulent Earth bioshell) of damaged, heavily discounted, distressed or sub-specification satellites, satellite pieces, various frozen fluids, and damaged vehicle parts is all affected by inconstant space weather generated by a fluctuating Sun (Miyake et al., 2017) during the most recently begun and still ongoing Solar Cycle (Schrijver, 2015; Baker and Lanzerotti, 2016). The absolute physical destiny for more than 5,000,000 kilograms of deliberately orbited materials is unknown because the mass is inherently unstable, subject to known/unknown and many still unpredictable changes during Earthly Geologic Time's passage—use, value and even to some extent function. Strictly speaking, it cannot yet be named "junk" because that label implies that it has some potential retrievability and normally Green-prompted re-use value; briefly, the term "waste" has an ontological ambiguity—it can be an excess, surplus, burden, resource—however, wise re-use of surplus is more appealing/thinkable to many persons than is excess material because if a surplus does not end in some ultimate use, then it becomes a disposal macro-problem which imposes economic and homeland planet bioshell burdens! Loosely scheduled for launch into outer space during 2017/2018, the 100 kilogram, 80 centimeter-diameter bubble-shaped hollow Earth-orbiting **KEO** satellite, carrying 100 DVD discs, a diamond casing enclosing a drop of Homo sapiens blood, and samples of Earth's air, seawater and soil, is slated to safely re-enter the Earth-bioshell, bobbing on the ocean's surface for a long time or stably sitting on a landscapes somewhere, about 50,000 years after its rocketing into outer space. **Figure 3.** Its return may be marked by a temporary supra-stratospheric aurora phenomenon, Earth's extant geomagnetic conditions permitting (Owens et al., 2017). **KEO** was conceived during 1994 by space artist Jean-Marc Philippe (1939-2008). Five hundred centuries hence, **KEO** might be dubbed a rudimentary physicalization of the "resurrection paradigm" espoused by some avant-garde biodiversity specialists (Franks, 2008).

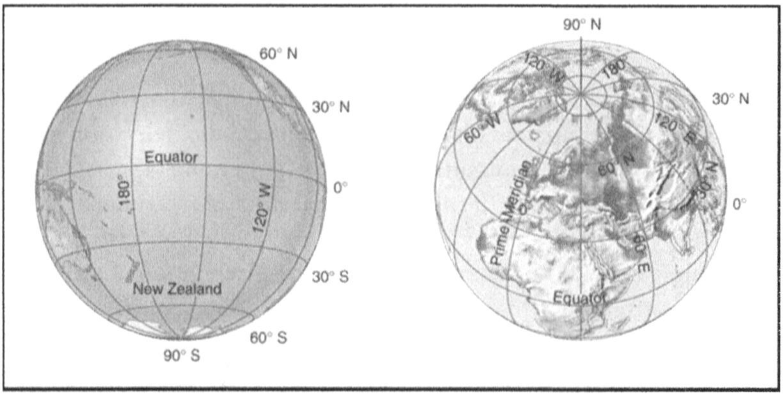

Figure 3. A map of our Earth's two hemispheres—left-side: its water hemisphere, right-side: its land hemisphere. (Google Image helpfully modified by JMH.)

Some popularly famous outer space probes composed of contrived physical materials exiting our Solar System, such as **Voyager 1** and **2**, **Pioneer 10** and **11** were partly designed by Carl Sagan (1934-1996) and his socially like-minded allies and immediate compatible compatriots to perform as Homo sapiens-revealing "spoor" [***Voyager Golden Records***] for space probe-intercepting Aliens. [Team Geographos, in Chapter Four of ***Sci-Fi Macro-Imagineering***

(2017) counseled strongly against sending any messages to Aliens anywhere in the known Universe!]. These manufactured macro-objects may carry dormant, but still viable non-inventoried biota, even though it was never any rocket launcher's intention to instigate any extra-Earth Astronomical Infestation Effect. Skilled and observant Alien trackers could possibly deduce and geometrically project from whence the spoor-littered trail created by these drifting, non-functional, technical material and parts-shedding machines likely originated in the Milky Way Galaxy. Any human or group of humans, encapsulated by a very sturdy spacecraft, accelerating steadily at a comfortable rate of 9.81 meters per second per second, could reach 99% of the distance to the universal micro-wave background's interior "edge" well within a normal human life-time (Anderson et al., 2016; Kwan et al., 2010)!] The nowadays little-practiced human art of tracking living as well as mechanicals—afoot, not via drone or other aerial or satellite remote sensors—within the Earth's bioshell may have been the very first implementation of Science during Homo sapiens' cultural development (Liebenberg, 2006); it can, therefore, be safely supposed that ultra-mobile Alien spationauts, untainted by misoneism, will have followed a similar course of biological-intellectual cultural attainment. Any mindless invasive biota aboard the interstellar **Voyager** spacecrafts, 40,000 to 200,000 years hence, if Homo sapiens no longer existed or had retrogressed intellectually, or the whole Earth had been disrupted by some natural or unnatural catastrophe or by far-seeing macro-project planners, would then have the status of homeless orphaned wanderers, origin only barely deducible by those wielding super-advanced surveillance, chemographical analysis and space-time traveling technologies!

Coherent, powerful energies generated and comprehensively broadcast in a widespread manner by humans since the invention of radio, television and radar, traveling at light-speed, will announce (or already has) Homo sapiens' presence to all attentive Aliens equipped with appropriate detectors and recorders. Sometimes such broadcasts—not really intended for Alien reception, but having that effect inevitably—are collimated beams aimed at specific places in our Earth's sky. For humans, radar astronomy will increase in usefulness because of its obvious indispensable employment in assessing the true asteroid Earth-impact hazard as well as scanning the surfaces of other Solar System planets. There is a quantitative and qualitative uncertainty of risk to people associated with energetic broadcasts, whether focused or unfocused in nature, that may be observed by Aliens, and this risk is preliminarily rated and categorized by the San Marino Scale (Almar and Shuch, 2007; Gertz, 2017). Possibly there may be Alien spationauts with tendencies for interstellar predation (Cockell and Lee, 2002). It would seem wise that species-wide human-Alien theoretical social relational policy, at least in public at a global scale, be silent on the subject of capital goods and homely comestibles. [This foreboding topic is dealt with comprehensively in Chapter Four of ***Sci-Fi Macro-Imagineering*** (2017).] Demographers forecast eight billion human beings could be present by 2024 but, confusingly, they often report too a trend amongst Homo sapiens for future stable or even reduced numbers. Since recognized innovation by young workers and entrepreneurs may decline, then a period of global Homo sapiens economic decline may ensue unless AI robotics replaces unborn thinkers and doers; too, it is wise to recall that the Latin "poplo", today present as a prefix in our English-language term "population", means "army"! AI robot replacement of humans in all respects can happen only if people format themselves based on AI robots imitation; but any sufficiently large dataset contains arbitrary—hence spurious—correlations so that AI Robot algorithms must always be flawed (Longo, 2017). Equivalency embodied in Nature-designed bone and flesh or, alternatively, in purely artificial materials (Pulido et al., 2017; Yeomans, 2017)! Even potentially, actual interstellar trade between Aliens and humans is a very improbable commercial process-event mainly because of the enormous interstellar distances of space separating their respective stellar systems or rogue planet civilizations. However, since Team Geographos, in concert with many others researchers, surmise that Alien macro-artifacts may be encountered by human or AI robotic space travelers eventually, it also is logically presumable that Aliens will indulge in some form of archaeology centered on non-living, Homo sapiens-fabricated space debris.

Sealed-capsule Homo sapiens Projected Biotic Interstellar Artworks

Extra-terrestrial outer space exploration is the greatest voyage of discovery ever undertaken by Homo sapiens. Using a benign form of spatial exploitation, artists have been at the forefront of interplanetary and interstellar space exploration since the commencement of our species' Space Age (Raitt, 2007). For example, the Soviet spationaut Alexei Leonov (born 1934), the first man to conduct extra-vehicular space-suited activity, also used colored pencils on 18 March 1965 to form a cheerful, simple line-drawing whilst orbiting in outer space, and possibly inaugurating the genre of Art made in outer space! Larry Burgess's ***Boundless Cubic Lunar Aperture***, launched into Earth orbit from the U.S.A. in 1989, was the "...first official non-scientific payload put into space" (Gefter, 2007, page 52). In common English-language parlance, "payload" is a synonym for "carrying capacity", a consistent scientific/engineering measure of how much X an inanimate package Y could transport over a given distance (Sayre, 2008). Our world's first printed material—micro-reinforced concrete (Cathcart, 2011)—was employed to construct a pedestrian footbridge serving human walkers, 12 meters in length by 1.75 meters in width, quickly erected in Madrid, Spain during 2017. Meantime, the sculptor Eyal Geyer, with the NASA's assistance, on 10 February 2017 exhibited, ***#Laugh***, a made-in-place—aboard the

International Space Station—utilizing a confidential, specially-developed material solidly printed sonification of selected crowdsourced—a word first popularized in ***Wired*** magazine—smartphone application human laughs because laughter often provokes more laughter! Geyer retains hopes his sculpture will stay in outer space forever. His event-statement hand-out: "A laugh star floating in space, above our heads, is my attempt to create a contemporary metaphor for the hanging 'Sword of Damocles', a reminder that the beauty of human life is so fragile". [Team Geographos can only fuzzily macro-imagine what this cacophonous world's interstellar anthrophony transmissions would appear if it was physicalized as an Alien printed collected artwork!] With regard to humans exclusively, the concept of "carrying capacity" is profoundly over-applied to the extant Earth-bioshell as the notion of Homo sapiens' "ecological footprint". Radical Green philosophers too often use this vague, almost pseudo-Science concept—related to the indescribably silly notion of Nature's Balance mythology—which was, years ago, dismissed as an inescapable self-contradiction (Frazier, 1997). Almost certainly humanity's bioshell will be more degraded as remaining wildlife habitats, free-flowing freshwater resources, and safely breathable air are hastily, and rather unwisely Team Geographos feels, dedicated to human civilization's cropped land-and-freshwater appropriation to sustain the radical Green farmer-led bio-fuel craze (Haber, 2007); in fact, humans may still be courting a planetary "ecocide" event-process (Teclaff, 1994) as bad as befouling fossil fuel use, after costly exhumation, because of the global popularity of this partly erroneous, overly-propagandized radical Green idea! It would seem the aversive "Green Revolution" of the 1970s that produced more food for people and their herds of domesticated livestock simply established the pre-conditions for the ongoing biofuels-developing, ecosystem-state social Green lunacy. Even more bizarre is that different, normal Green R&D routes to discretionary and base-load energy use are becoming available and competitive during the 21st Century!

Whilst Science provides an interesting intellectual framework for Art, both rational and irrational artworks can arouse and express human scientific intuition, as entertainingly revealed in detail by Art historian Andreas Broeckmann's ***Machine Art in the Twentieth Century*** (2016). A remarkable visual reminder that Art is central to our bioshell can be illustrated by obscuring the first and last letters of "earth". Those scientists who have examined and studied our planet's changes through at least 4.5 billion years of Earth's Geologic Time have postulated that Earth's life may have started in, and first flourished on, moist clay deposits, or life may have finally landed on Earth after descending from our Universe, an unexplored spacial vastness thoroughly proved to be a cryogenic habitat for dormant microbial life-forms! Other scientists have good reason to deduce that Earth's earliest life-forms sequester themselves inside gas-filled bubbles at the seashore 3.2 billion years ago to avoid extinction caused by Earth's then intense

UV radiation bath (Homann et al., 2015). Our Universe seems to have the great possibility of being a meta-ecology spatial realm. In other words, as of early-2017 AD, Earth must be considered a complex globular local Solar System "maternal figurine" housing plant and animal life in a thin shell-like planet covering named bioshell, or incorrectly as "biosphere". Indeed, its Second Nature characterization by those trying to sell the concept of the ceno-Anthropocene must mean, logically, that ours is a world that now has no vertical parts of its crust or horizontal geographical regions atop the deepest Earth-crust, closest to the Moho stratum's mantle-crust boundary, that can be left undisturbed or to grow wild. Since 1936, Earthly microbes were first collected aerially from high altitude by balloons, aircraft and even meteorological sounding rockets. Japanese artist Makoto Azuma's ***Exobiotanica Botanical Space Flight*** conveyed a 50 year-old bonsai tree and an accompanying arrangement of flowers to Earth's upper stratosphere during 2014 in order to apprehend the appearance of its living payload when cast against the bluish Earth background. Both Science and Art look to the future and seek truth. Organisms have long been underestimated in terms of their survival under adverse environmental conditions and, recently, a green algal strain [CCCryo 101-99 of Sphaerocystis sp.] and blue-green algal strain [cyanobacterium Nostoc sp, CCCyro 213-06] endured, undamaged, 1.5 years of caged exposure to nearest outer space, latched onto the outside of the International Space Station! **Figure 4** (Google Image, below.)

The technical, laboratory culture of some living cells, known and unknown, named and unnamed, redefines "biota" as life becomes disembodied and widely distributed anthropogenically on other celestial bodies, first in our Solar System, then perhaps elsewhere. Artworks composed mainly or wholly of biological displays are ever more commonly encountered in ateliers, commerce's art-galleries, and avant-garde museums of Modern Art. Genetic Engineering's R&D is an

Art-driving concept that permits astute artists to freely create future life-based artworks that are original, unique, and endlessly reproducible and, furthermore, that near-term future life-based artworks may stem from the results of continually progressing commercial Bio-Engineering. Stephen Wilson's ***Information Arts: Intersections of Art, Science, and Technology*** (2002) tastefully tout that 21st Century macro-imaginative concept whilst being a significant step beyond the previous quintessential basic summary in ***Art and Physics: Parallel Visions in Space, Time, and Light*** (1991) provided by surgeon-artist Leonard Shlain (1937-2009). Since DNA was first introduced as a secure and decodable data-storage strategy that means humans must not be careless in what is thrown away by our civilization as "junk" or intentionally sent as "mail" to Aliens (Erlich and Zielinski, 2017). If any future direct interstellar contact attempts are (unwisely) completed by Homo sapiens, then aesthetic principles might have been inserted into those anthropic intellectual and technological efforts.

An unusual Panspermic Astrobiology viewpoint is recounted herein, asserting that known life, chiefly in the form of sack-like living biotic artworks (cellular), may become a significant near-term means of proliferating and transporting our planet's particular life-forms throughout interstellar space adjacent to our Solar System. [Team Geographos affirms that, so far, the exact number of species inhabiting Earth's bioshell is not ascertainable (Sukumaran and Knowles, 2017) and, as well, that Earthly DNA has been discovered to have the property of being a reliable, rewritable random-access data-storage system (Yazdi et al., 2015). Earth's bioshell may be home to as many as one trillion microbial species whilst the total DNA of all life-forms within our planet's bioshell is almost inestimably large (Landenmark et al., 2015).] It was an inquisitive, forward-thinking artist, Joe Davis, who in 1988 was the first person to "...map the ones and zeros of digital data onto the four base pairs of [Escherichia coli] DNA" (Extance, 2016, page 23). Such artworks could penetrate the constraining bioshell "bubble" to which our species was totally restricted until 4 October 1957. For instance, the Polish artist Zbigniew Oksiuta's imagined ***Space Garden*** (2002) daringly opted for an air-filled hollow anthropogenic orb, fashioned of biological polymers, cast above our planet's atmosphere into outer space and functioning as a bioreactor housing transgenic plant experiments (Oksiuta, 2007). Similarly, Tomas Saraceno, during 2006, proposed a flying garden (Anon., 2006). In both examples, Team Geographos was struck by the idea's similarity to one voiced many years earlier by American rocketry pioneer Robert Hutchings Goddard (1882-1945), when he foresaw people-transporting interstellar arks: "It has long been known that protoplasm can remain inanimate for great periods of time, and can also withstand great cold, if in the granular state" (Goddard and Pendray, 1970, pages 1611-1612; Armstrong, 2017). Space-faring mini-bioshells, space-suits for non-humans, of the kind vividly envisioned by Zbigniew Oksiuta could be fabricated in outer space polymerization using epoxy, or enclosed by polyethylene composites, which are materials strong enough to sheathe and protect spaceships from micro-meteoroids and, approximately, 15% of all impinging Milky Way Galaxy cosmic radiations. By harnessing micro-gravitation, temperature variations, hard-vacuum, UV and VUV, and cosmic rays, Oksiuta refreshingly predicts that, eventually and most probably during our 21st Century, Homo sapiens-selected strains of DNA will be deliberately encapsulated inside outer space-lodged bioreactors for the express purpose of promoting autonomous development of new forms of life! One kilogram of the bacterium Escherichia coli DNA could store all of our world's Homo sapiens-selected data—more than an elephantine memory in a physically small-sized packet! ARTSAT1: INVADER, a cubic reporting satellite 10 centimeters on a side, inserted into Earth-orbit on 28 February 2014, issues solar-powered broadcasts of synthetic voices, music and poetry to ground stations, perhaps the first independent use of a satellite as a personal media (Gracie, 2016). ARTSAT1: INVADER poses the obvious future introduction of cellular automata into anthropogenic baubles moving through outer space as tools for creative pattern-generation (Shin, 2016). Ever since the microscope was first used after its inversion from telescopic studies, it has become evident that microbes commenced the Earth-bioshell's formation as well as its continuation as a possibly unique entity in our Solar System (Falkowski, 2015)! Interestingly, Escherichia coli can survive hydrostatic pressures of up to, and possibly exceeding, about 19,739 standard Earth atmospheres (Vanlint et al., 2011)! Most computer chips cease to function reliably at approximately 250^0 Celsius. Last year the NASA reported that it had, through costly taxpayer-funded but government-mediated R&D, achieved a computer chip that may only now be barely capable of enduring its pervading caustic chemical environment, incessant $+460^0$ Celsius heat, and about 94 atmosphere pressure of the mostly carbon dioxide gas-enshrouded planet Venus (Neudeck et al., 2016).

Irritating the Elephant in the "Room": Known Universe as Homo sapiens' Design Studio

Some physical aspects of such remote and unattended gardening endeavors can be particularly conducive to independent evolution. For instance, spun-off ***SEEDS: Synergizing Earth's Evolutionary Development Spacewards*** (1997) by Arthur Woods and Oksiuta's ***Space Garden*** are subject to space weather, which is more than a mere annoyance for all living creatures, big or small. Space plasma can kill macroscopic life-forms and can mutate and nourish life, yet it has been demonstrated that better-than-elephantine memory technology utilizing living organisms is of much greater potential than any of the existing counterparts to render a service of inheriting big data. Such corporeally encoded data can be preserved with insignificant corruption for hundreds to thousands of years. Zbigniew Oksiuta has yet to announce any intention of replicating in outer space the valuable seed-filled safe-deposit vault near Longyearbyen on Spitzbergen, one of the Svalbard islands that Norway's Global Crop Diversity Trust has operated since 26 February 2008. AD 2019 will mark the 160th anniversary of the publication of Charles Darwin's best known book on the origin of Earth's many plant and animal species. Oksiuta has no announced plans to honor Darwin (1809-1882) by a copy-cat repetition of 1989-1999s ***Millennium Rock*** episode when a dolerite chunk was expensively lifted by fiery rocket into an Earth orbit as an arrogant American's public-relations stunt supposedly meant to mark the Darwin book's 150th anniversary (Parker, 1999)! Far-distant Alien radio- and neutrino-delivered artwork diagrams have been "decoded" by some apparently very gifted human artists. Examples: (I) Jonathan Keats' ***The First Intergalactic Art Exposition*** (31 July 2006 to 14 January 2007) at the Magnes Gallery in San Francisco, State of California, U.S.A.); (II) ***Tom Na H-iu*** exhibited at the 2006 Singapore Biennale by its creator, Mariko Mori. Other strictly biological artworks may already have been assembled specifically, and broadcast deliberately, by gregarious intelligent Alien civilizations, extant as well as extinct, as the only realistic means to materially perpetuate every civilization's independent cultural achievements within the enormous volume, and violence, of our known Universe. The only Alien or Homo sapiens instrumentalist activity that could possibly use technology totally is material mass destruction. [METI International's (established 2016; www.meti.org) publically-announced plan for near-term future Messaging of Extraterrestrial Intelligence—"Hello Aliens, We're Tasty Earthlings and We Dwell Nearby!"—really ought to be postponed, indefinitely (Vakoch, 2016)!] Is this why so few scientists, and only a miniscule social-set of artists, have openly speculated the known Universe may have commenced with a "Big Bang", and might terminate with a "Big Crunch" initiated artificially? This potentially productive facet of academic and popular Astrobiology—literally the ostracized Elephant in the Room—has remained mostly unexamined! Few Astrobiology textbooks make any reference to "Art" or "artworks". Therefore, Team Geographos feels compelled to gently suggest that, in future, scientifically studying such hitherto unexamined astrophysical aspects of Alien intelligent activity can become a useful widening of our present-day modest SETI-related efforts. In a sense, these remotely detectable Milky Way Galaxy artistic Alien things—that is, actual material objects and/or energies—will "talk" as common spoor "speaks" to trained and artful human trackers! During July 2014, the **UKube-1** Earth satellite was placed in orbit, supposedly bearing the first-ever satellite artwork etched into its side—asserted by public relations spokespersons as a kind of "satellite tattoo"—created by the Los Angeles, State of California-based firm *iam8bit Art Gallery* [www.https://iam8bit.com]. The so-called tattoo was designed as an invitation to passing Aliens to stop and utilize **UKube-1** as an electrical charging device for their communications, spacecraft and other relevant devices! Oh, what a predictably politically correct radical Green sentiment!

Earth's Bioshell: Our Solar System's Primary Natural Chia Pet

Peter Ward's ***Future Evolution: An Illuminated History of Life to Come*** (2001) forecast Earthly life's shapes and activities most adventurously, if not futuristically. However, he regrettably neglected to elucidate any possible future for Earth's life and/or its changes over Geologic Time in any way related to nearby outer space. A few Europeans during the 19th Century, most notably the geoscientist Carl Bernhard von Cotta (1808-1879) and botanist Franz Unger (1800-1870), confidently predicted some forms of Earth's life might evolve to replace Homo sapiens as the dominant bioshell factor. Early 21st Century human robotics experts programmed their intellectually inferior (so far) metal and plastic wards to be evasive in their real-world physical and intellectual interactive behaviors with humans. But, at the beginning of the 20th Century, chemist Svante Arrhenius (1859-1927) proposed a theory of Panspermia, the life transfer from any world with a bioshell to another world, barren or with bioshell, which has been resurrected in the fervid context of contemporary Astrobiology—that is, the known Universe's meta-ecology; Arrhenius suspected that individual spores might be propelled through interplanetary and interstellar space by solar radiation pressure. A particularly speculative version of panspermia, initially proposed in 1973 by biologists Francis Crick (1916-2004) and Leslie E. Orgel (1927-2007), is the idea of "Directed Panspermia". Homo sapiens' Space Age had furnished investigators with data that, at least, 100 metric tonnes of materials fall onto our planet daily, and some of it may be biological in origin! Some scientists and non-scientists have speculated that passing comets, or those that collided with Earth, seeded Earth with life; the unusual green-colored comet 45P/Honda-Mrkos-Pajdusakova zoomed past Earth on 11 February 2017, missing our homeworld by about 32 times the average Earth-Moon interval! And, a mass exceeding 5,000,000 kilograms has already been sent by us into outer space from Earth and to the probably abiotic surfaces of other celestial bodies within our Solar System. It is likely that more than 200,000 kilograms of human-designed and fabricated macro-artifacts currently rest atop other Solar System planetary crusts, on several asteroids, and even been smashed into comets according to macro-plan! This outstanding 20th and 21st Century extra-Earth bioshell accomplishment remarkably remained unrecognized by any of the nearly fifty chapter contributors to Dr. Barney Worf's informative ***Handbook on Geographies of Technology*** (2017). Of course, even today future terraformers do plan to ship entirely Earth-sourced living organisms into Mars to transform the rugged and trough-riven dryland surface of the Red Planet, literally coloring it green with verdant familiar vegetation after its human-led conversion to this Solar System's secondary planet-sized Chia Pet! Considering the prospects of human colonization of Mars, then application of perfected bio-mining for extraction of useful elements, or for anthropic formation of soil, would be rewarding (Cockell, 2010). Humans may have already deposited a few organisms unrepresentative of our Earth's total global Nature in Mars; proponents of Mars's settlement have endorsed Synthetic Biology as a means to successfully colonize (Verseux et al., 2016) and some have endorsed the outdoor use of synthetic biota within Earth's bioshell (Sole et al., 2015; Hutchinson et al., 2016).

Our Earthly bioshell has been a geographical zone of extensive and intensive chemical activity for billions of years. Its base, the Moho stratum, is our planet's crust where that unit of structurally heterogeneous collection of tectonic plates—the total number enumerated varies in published geoscientific reports—interfaces with the Earth-mantle, and its cap is formed by the interface of Earth's so-called biota-free upper atmosphere with outer space. Commercial airliners utilize the stratosphere to about 16,000 meters and the International Space Station orbits far above; the atmosphere layer between is unused by human beings. That is, the air layer shell from altitude 17,000 meters to 51,000 meters is presently devoid of functioning anthropogenic infrastructures although World View Enterprises, a private-sector company based in Tucson, State of Arizona, plans to launch tourists in its **Stratollite** balloons by AD 2018 to reach the edge of outer space [30,000 meters], offering **Stratollite** riders elevated views of, among other things, our planet's visible curvature. Micro-organisms artificially projected into outer space's hard vacuum are quickly heated by the infra-red spectrum of sunlight and then they sometimes die as they explode. Nevertheless, our singular bioshell is, indisputably a geophysical "greenhouse" [another Physics misnomer, like Biology's biosphere] that is increasingly under the influence of Homo sapiens through enhanced atmospheric effects, including unnatural outdoor lighting, emission of gases of various kinds, even the remains, as ashes, of cremated human beings (MacMartin et al., 2013)! Incorporated during 2016, MesoLoft in Lexington, State of Kentucky, U.S.A. offers commercially to lift a departed loved one/hated one contained within a [recycled?] paper payload [a Nature-recyclable ash-holding urn] to nearly 23 kilometers via a hot-air paper balloon—almost identical to the Japanese Kamifusen described by Fukumori (2017)—for particle free-fall after release. MesoLoft offers balloon launches from the U.S.A. states of Colorado, New Mexico and Indiana. Rubber weather balloons burst at 30,000 meters above sea-level (Cullis et al., 2017). At its official WWW website, www.mesoloft.com/science, the company's online sales promotion literature offers that "...ashes can attract water vapor deposition and increase in size, to the point that they can serve as nuclei for [water] droplet formation". As of AD mid-2017 just two Earth-atmosphere mini-aerial weather and climate modification efforts have been conducted, so far: (I) in 2011 University of California researchers sprayed salt granules into the open air over the Pacific Ocean near San Diego and (II) in the Russian Federation, experimenters dispersed sulfate particles from a helicopter and automobile during 2009. However, next year [2018], Harvard University scientists Dr. David Keith and Dr. Frank Keutsch intend to launch, from a place near Tucson, State of Arizona, balloons named **StratoCruiser**, which would spray fine mists of sulfur dioxide, alumina or calcium into the stratosphere to test the real-world reflectivity of such granules in order to accurately assess the usefulness of such particles for any regional or global Geo-Engineering macro-projects! American geographer Dr. Stanley D. Brunn penned a revealing poem, ***Don't Let the Rains Come***, dated 10 February 2017, to memorialize MesoLoft's timid ambitions in a unique fashion! Approved by the U.S.A.'s Environmental Protection Agency during 1997, cloud-seeding of this sort, done with the powdery aerosols of passed humans, will possibly, according to the very wise Dr. Stanley D. Brunn, geopolitically "...raise [currently religiously over-heated?] inter-faith discourse to entirely new heights":

> "In a colorful hot air balloon with a video camera attached.
>
> Upwards of...25,000 meters...your ashes will quietly go
>
> Then opened in return to earth in clouds, rain or fresh fallen snow.
>
> The camera will send friends photos of earthly ashes on this heavenly trip
>
> To share with others on Facebook, cell phones and the Internet."

Human influence is so pervasive and densely represented in some places that leading fashion garment designers have created adaptive clothing tailored to the anthropogenic new global warming conditions imposed on Homo sapiens everywhere. The items of apparel are not much different than proposed terraformed Mars counter-pressure suits as well as present-day and future embodiments of space-suits (Hart, 2013; Cadogan, 2015), which are human-shaped garments fabricated to maintain pressurized air for a comfortable living person, not a body-bag or cadaver pouch isolating an unfeeling deceased person! Thus, obvious and scrutinized global climate change has become a professional métier for geoscientists and artists of various persuasions. In other words, by using vast arrays of technologies, humans are expanding the volume of artificial climates generated for clothed individuals and structurally delicate greenhouses. As macro-imagineers, Team Geographos has proposed Earth's complete enclosure with an anthropogenic plastic/concrete slab cladding costume that will preserve, protect and maintain present-day climate regimes to a certain extent, as well as preclude forever any form (theoretical as well as real-world) future human homelessness within Earth's roof-contained bioshell (Cathcart and Cirkovic, 2006, pages 151-174)! Such an enclosure can only be termed a true para-terraformation mega-project except how does one make Earth more...Earthly? By conversion to Mr. Oswald Wiener's "Bio-Adapter", an enlargeable machine for human living that is far more inclusive, adaptive and comprehensive than any mechanical thing portrayed by Thomas Carpentier in Ellen Lupton's ***Beautiful Users: Designing for People*** (2015) and the "Phytotron" design of Frits Warmolt Went (1903-1990) historically examined by David P.D. Munns' ***Engineering the Environment: Phytotrons and the Quest for Climate Control in the Cold War*** (2017)!

Whilst still an uncapped planet, our widespread Earth-based extractive industries—those that gather exhumed fossil fuels, ores and mined aggregates—have provided some sources of energy, other than the Sun, as well as other raw materials required to instigate global warming. Global warming stimulates construction of plant conservatories to ensure biodiversity and continues to alter the distribution and intensity of outdoor photosynthesis worldwide. The carbon dioxide gas exhaled by Homo sapiens amounts to about 600,000,000+ metric tonnes per year, or about 12% of fossil fuel combustion annually (Prairie and Duarte, 2007). Statistically, each Earthly year, every living human—there are now in excess of 7.3 billion persons alive in 2017—will inhale approximately three particles shed by a terminal meteoroid during its rapid descent through our air, and there are many little-known organic materials that constitute our wind-moved air's burden (Goldstein and Galbally, 2007)! In other words, the biological identity and technologized culture of Homo sapiens occurs presently at a "terminal" in the Earth-bioshell; that "terminal" can be both an ending and an interfacial zone. Since some of those person-inhaled descended aerosols may be living, dormant or dead biota, an Alien "interface" with the Universe is already an environmental fact and remains, so far, a generally unacknowledged everyday other-worldly, but real, happening that is remarkably akin to Alien-to-Homo sapiens METI (Yokoo and Oshima, 1979); perhaps humanity is being progressively cultivated, by technically-advanced, Alien-formatted soft automata programmed explicitly for the care and feeding of previously discovered/implanted ancestral Homo sapiens? Only the continued advancement by humans of Biotechnology and Genetic Engineering holds the promise of the definitive "yes" or "no" to this questionable proposition of Alien-to-Homo sapiens genetic METI. And, Alien life-forms need not have been extra-Solar System, speculatively they might have been a prior but now extinct species (on Mars, for instance) (Wright, 2017).

Bedrock Geoscience: Foundation for Soft Michani-panspermia?

Built in the U.S.A.'s mostly desert State of Arizona, and first occupied by humans during 1991, **Biosphere-2** was partly composed of concrete—in effect that single type of concocted material (comprised of freshwater, sand, aggregate and cement) was used to represent the many different life-nourishing materials composing the Earth-crust (Kirkby, 2016; Anker, 2005)! The effort was a failed USD$ 50 million Earth-bioshell ecosystem experiment. And, R&D in related necessary technologies has, since **Biosphere-2**'s ignominious emergency shutdown, far surpassed anything available to researchers then: entomologists recently have documented that some Earthly food plant pollinators, a species of bees, that nest in bee-excavated holes in sandstone rock—resembling the rebar-less concrete sections of **Biosphere-2**—in wild xeric ecosystems of America's Southwest certainly exist (Orr et al., 2016). Bees, approximately 3.2 trillion in number globally are trending however to become a near-term future endangered species, perhaps even a rarity in certain major agricultural regions. So, rising to a farm industry challenge to profitability, human roboticists have suggested that robo-bees as natural bee assistants, or even alone, might do the unheralded job of seasonal Earthly plant pollination someday in order to ensure that the 35% of food crops needing bee pollinations are maintained properly by supplementary robo-bee assistance to Nature's uncontrollable bees (Chechetka et al., 2017)! Zbigniew Oksiuta, during 1995, received the Award for German Architects when his artistic focus was then solely on concrete architecture. At about the moment the Chia Pet was first marketed in the U.S.A., during 1982, Richard Dawkins daydreamed in ***The Extended Phenotype*** on the prospect of a Martian geoscientist trying to "make sense" of remotely apprehended human life-forms and technologies. Team Geographos' proposed slow-moving spacecraft—that is, Oksiuta's artistically original ***Galactic Space Garden*** as we have agreed to retain it for its charm—will facilitate a directed michani-panspermia macro-project to deliberately move some living organisms and a DNA record of our Homo sapiens civilization through interstellar space and not only bombard targeted Aliens with our distasteful popular culture via radio and television transmissions as Dawkins feared. Certainly, smart people can best in work performance mere robo-bees! An informed Astrobiology guesstimate reveals that the "...total amount of life in a habitat of finite duration, such as the Solar System, may be measured in terms of time-integrated biomass. In these terms, the potential amount of future life about the Main Sequence Sun can be 10^{34} kilogram-years, largely exceeding the 10^{24} kilogram-years of past terrestrial life" (Mautner, 2005, page 167).

Will Earthlings, unassuaged by radical Green dead-end nostrums, established within an influential 21st Century globalized grassroots organization, play a marked role in furthering the perfection of uninhabited anthropo-kinetic directed biotic artworks that are propelled and directed to leave our Solar System? Effectively, such anthropogenic objects containing specially chosen life-forms would become talking-headstones, far more complete than anything yet macro-imagineered and voiced by Gary Gumpert's ***Talking Tombstones and Other Tales of the Media Age*** (1987) that exposed some of the means by which common human cultural values and universal human beliefs were then being altered by media technology innovations. Oddly, Robert Barrow's awarded U.S.A. Patent 7,089,495 for a video-enhanced grave marker involving a flat LCD touch-screen installed within a hollowed headstone, was not officially registered until seventeen years later! During Earth's ancient Geologic Times, as subjectively determined by emotional humans of course, the Greeks of olden-times told a myth about Gorgons which involved three winged female monsters, one of whom had the power to convert a person to stone with a mere glance. In today's English-language parlance "to gorgonize" a person is to metaphorically petrify someone with an intimidating glare. [In the Hollywood, State of California, U.S.A., Universal Pictures issued ***The Monolith Monsters*** during 1957, at the start of humanity's Space Age, the science-fiction equivalent of the ugly Gorgon of Greek mythological fame, with fallen shattered rock meteorites as cinematic stand-in for the Gorgon.] Geoscientists and others continue to seek means to manufacture stone by the application of extant and future technologies. Some bold geoscientists, following the history of artificial stone creation, have not been intimidated by the alleged looming threat to all life-forms of increasing our air's content of carbon dioxide gas. Such courageous-in-the-face-of-radical Green-inspired public opinion are approaching an economic solution to the macro-problem of providing a long-term technological fix to the very real build-up of CO_2 caused by Homo sapiens' instigated carbonation reactions after direct extraction from the ambient air, which has the potential to alleviate some unwanted or undesired climate regime changes (Kriegler et al., 2016). It is also possible that humanity could come to think, collectively, that the exportation of greenhouse gas and water-vapor from the Earth in Zbigniew Oksiuta-inspired encapsulations of his ***Galactic Space Garden*** mini-globes—a kind of mobile sack designed specifically to hold things conveniently for inter-stellar transports is workable and useful. Traditionally, "sack" was the last word spoken at the Tower of Babel before confusion of humanity's spoken tongues.

For millennia—yet uncountable Geologic Time years after bees had become extant—humans dug and piled only natural Earth-crust rocks. Archaeologists have reported that artificial basalt was manufactured in ancient Mesopotamia, reputed construction site of the Tower of Babel. In the UK, Victorians were familiar with the interesting properties of durable mock rock surface formations artistically shaped in urban public parks that were composed of Pulhamite and Coade Stone (van Lemmen, 2006). The constancy of factory rocks—blended concretes called geopolymers—featured in many attempts undertaken since 1945 to immobilize dangerously anti-biota high-level radioactive wastes. At tempera-ture less than 100^0 Celsius, geopolymers can poly-condense like the organic polymers that fascinate Zbigniew Oksiuta, Raj J. Kaul and many artists. They are UV and IR resistant and, in pure composition, could house a biological culture, which is a colony of cells grown on a nutrient medium under controlled condi-tions in an incubator. Typically a thin stratum of jelly comprised of agar, sugar, freshwater, and other materials provides basic food-stock for the cells, nourish-ing the colony. Dr. David J. Sheppard (1979) asserted that Sun-orbiting human space settlements, and even interplanetary spacecraft and interstellar world-ships ought to be formed exclusively of outer space manufactured blended concrete. Dr. Sheppard understood that opaque concrete is rather inert and would not be too vulnerable to impinging space weather or atomic oxygen erosion that can readily occur to a satellite in low-Earth orbit. It can be formulated to be structurally thin, to incorporate flexible glass micro-beads, latex, or acrylic fortifiers; there is no chemical or physical reason whatsoever that this type of architectural shell, a "bubble" cover, cannot be well-larded with biota-nutrients (Bartlett et al., 2017). [For example, SEE: U.S.A. Patent 5,948,474 awarded 7 September 1999 to Norman Andrew Tiller for "Bio-Active Rock and Method for Making the Same".] On 18 February 2017, the methicillin-resistant infectious pathogen staphylococcus aureus [the human-feared MRSA plague, (Crossley et al., 1979)] was sent to the International Space State so that biologist could observe and document gene expression and mutations, if any, occurring in the micro-gravity of low-Earth orbit. Biologists and other scientists, internationally, have urged the advance of Genetic Engineering aimed at the creation of "living spacecraft" (Hansson, 1996; Armstrong, 2017). Team Geographos honestly expects that improved geopolymeric cements, with considerable translucency, maybe become an intentional resource of stored nutrients for live organisms enclosed by innumerable mobilized robotic and/or Homo sapiens-fabricated spherical Zbigniew Oksiuta-inspired ***Galactic Space Garden*** "bubbles", almost "maternal figurines" traveling through interstellar space, perhaps the ultimate inside-out version of Earth's commercially marketed Chia Pet toy. Our readers ought to peruse Oksiuta's technology exploitation statement "Breeding the Future" to be found in the May-June 2009 issue of ***Architectural Design*** (79: 48-53).

Anthropogenic lithogenesis is a historically new event-process within our planetary homeland's bioshell. Nothing in the known Universe, or our most familiar stellar system, can last forever. But, manufactured rocks (concretes, ceramics and bricks) endure for a long period of Geologic Time, unless the planet containing the only human-known examples of life is shattered by a comet or enormous asteroid collision. What might be some of the really obvious results of the whole Earth's break-up, its terminal destruction as a planet-sized celestial body in this Solar System? Well-beyond, indeed objectively finalizing, Earth-crust changes caused by "human bioturbation" industrial activities (Zalasiewicz et al., 2014), R.B. Cathcart (1983) artfully outlined a feasible technical means, as well as the Astrobiologic results, of our planet's artificial terminal comminution—demolishment—by contemplative Homo sapiens leaderships. Concrete is the most used and familiar synthetic rock worldwide—approximately 4-6 cubic kilometers are mixed, poured each year! One of the main ingredients of Portland cement, the cohesive base for concrete, is limestone, life's by-product. Technogenic rock was first proposed as Fourth Class of rock-type at the turn of the 20th Century by American geologist Dr. James Ross Underwood (1927-2012), extending the commonly perceived igneous, metamorphic and sedimentary classes of rock-types.

[***Rocks: A Very Short History*** (2016) by Dr. Jan Zalasiewicz describes "Human-made rocks" in Chapter 8.] **Figure 5.**

Figure 5. Dr. James Ross Underwood, Jr. newspaper photograph taken sometime during the 1960s. (Image courtesy private collection of RBC.)

Dr. Underwood's practical instigative macro-proposal favoring establishment of an "Anthropic Rock" category recognized, at last, the pervading spread of humans and their industrial products. His reorganization came not a minute too soon, since the NASA, ESA and JAXA workers have since offered commercialized mega-proposals for the use of in situ resources, such as brick and the stereo-lithographic creation of vast expanses of photovoltaic carpets, by coating spationauts settling the Moon and Mars in the near-term future! Soon, artificial rock is likely to be generated in low-Earth orbit via focused radio waves. Dr. Underwood's timely theoretical innovation is a logical extension of the near-constant redefinition event-process that terms such as igneous, metamorphic and sedimentary have already undergone for several centuries because of Science's progress. Therefore, it is worth appreciating Norman Andrew Tiller's previously mentioned U.S.A. Patent!

Anthropogenic Traveling Vats of Various Earth Life-forms: Trajectories et cetera

Ordinary persons are coming to the conscious realization, as Mr. Oswald Wiener (born 1935) did during the early-1960s, that their bodies are planet-like, housing biomes of bacteria. Some biologists have alleged that each person is actually only 10% human—that is, each person's corporeal presence is composed of 10 trillion animal cells and 100 trillion bacteria, the latter contributing about 10% of the normal human body's dry-weight! Dr. Lynn Margulis (1938-2011) speculated that if humans continue to exist for a few million years more of Earth's Geologic Time, it is possible that some of the bacteria now residing (mostly) inside our bodies, such as micro-organisms that produce vitamin B12 in our intestines, may become integral parts of our own human cells (SEE also: Sagulenko et al.,

2017). Interestingly, plastic residues are ubiquitous in the Earth-bioshell and, quite obviously, exist too inside our normally functioning bodies (Calafat et al., 2008)! Dr. Harris Lewin, in his oral speech "The Earth Bio Genome Project" to Session 22: Final Plenary at the BioGenomics2017: Global Biodiversity Genome Conference, held at the Smithsonian National Museum of Natural History in Washington DC form 21 to 23 February 2017 announced that an international coterie of professional biologists intend during the 21st Century's coming years to sequence the DNA of all found and documented life-forms in Earth's bioshell.

Biotechnology's astounding progress inspired an R&D effort designated "Synthetic Biology"—now an accomplished fact, it is being considered by Earthly Geo-engineers as well as future Mars Terraformers as a technology today worthy of monetary laboratory investigation investments that will result in the widespread future outdoor use of anthropogenic organisms. 21st Century professionals working in the field of Astrobiology, and the world-public, have all been put on notice of some of the impending biotic developments, along with their generalized social and economic global impacts. Synthetic Biology undertakes gene synthesis and/or construction of artificial DNA (Gibbs, 2004). Assigned to J. Craig Venter Institute Incorporated on 31 May 2007, the U.S.A. Patent Application 20070122826 for a "Minimal Bacterial Genome" really kick-started Biotechnology's newest R&D track. Soon, it may become possible for industrialized Homo sapiens to summon, possess, and geographically widely utilize synthetic life at its perpetual beck-and-call! Perfection—supposing that is materially possible in this particular Universe—and widespread employment of synthetic organisms will, probably, result in Artificial Evolution. Defined most simply Artificial Evolution is the controlled micro-manipulation of Nature and laboratory provided genetic information, from one generation to the next, where the first genetic variation's shift is first Macro-Imagineered, follow by laboratory Macro-Engineered formation, and the second Nature's bioshell/artificial bubble encapsulation iteration is insured by Homo sapiens or, possibly, our biological species' AI robotic descendants!

Similar to the other reported outer space time-capsule macro-projects, our version of the ***Galactic Space Garden*** ought to be regarded as a form of insurance against the global catastrophic risks in the destruction of our particular civilization and/or the Earth's bioshell. Time Capsules are sealed deposits of cultural relics and recorded human knowledge intended for retrieval and opening at a selected target date, or accidentally found. The aforementioned **KEO** satellite best typifies this ultimate intention (Philippe, 1999). Team Geographos' speculated ***Galactic Space Garden*** will proceed on a planned trajectory that is a far lengthier route than a returning greenish celestial body such as Comet Lulin (C/2007 N3) moves through naturally—by the way, Comet Lulin's visible greenness stems from its cyanogen and diatomic carbon gas emissions. Either natural (asteroid impacts, super-volcanism, the ageism of Geologic Time; Dobretov, 2017) or anthropogenic (runaway Earthly "Greenhouse Effect" or global Nuclear Winter) catastrophes may, at some moment in our planet's future Geologic Time, interrupt the evolution of exposed life-forms and humanity's total civilization, and a desire for as wide as possible human presence in outer space is at least partially motivated by the desire to preserve as much as possible of the biological and cultural diversity of our planetary homeworld from existential contingencies. Zbigniew Oksiuta's inspiring and germinal **Galactic Space Garden** may, thus, serve an additional function as a repository for some of the important—however that may eventually be determined by its assemblers—Earth organism genomes. Francis Collins unabashedly asserted such biologically encoded data is ***The Language of God*** (2006); if Homo sapiens and/or AI Robots ever successfully foster truly Synthetic Life surely then that technological triumph seems to markedly undercut Francis Collins' rash statement!

Lofted into surrounding space from the Earth's bioshell, nudged into a trajectory that permits escape from our Solar System, Team Geographos' speculated living and mutating biota-globes, our adopted version of ***Galactic Space Garden***, will convey genetically encoded messages—pleas?, insults? threats? invitations?, dis-invitations? treats? jokes?, inconsequential chit-chat?—to technologically advanced Aliens and/or seed other star-orbiting corteges of sterilized or pre-biota planets. However, it is always possible some fastidious Aliens may view and classify captured Earth-originating artificial orbs as cosmic garbage (Gold, 1960) or, worse, as a provocation akin to personal and/or anti-Alien species insult, or blatantly aggressive germ warfare (Wolfe, 2002). More probably, this human action may be construed as a form of open and well-intentioned long-range physical communication which, in contrast to long-distance electromagnetic salty messaging—the infamous "Eat Doritos Now!" recounted in Chapter Four of Team Geographos' previous 2017 book ***Sci-Fi Macro-Imagineering***—leaves no doubt about the chemical and biological underpinnings of the sender (civilized Homo sapiens). Aliens might even use a still-functioning found ***Galactic Space Garden*** as a popular form of cherished collectable and display Art—in essence, an Alien fandom for our species! It is possible that a new Alien religion may come into vogue after our anthropogenic object becomes a studio, museum or parkland-conserved showpiece!

Earth Life's Pre-Space Age Trap Escape: Nature's Very Own Interstellar Panspermia?

Freidrich Ludwig Emil Diels (1874-1945), a German botanist who, more than a century ago, was the first modern-day scientist to notice as well as to document Earthly life's colonization of rocks (Diels, 1914). Dr. William M. Napier demonstrated conclusively that life-transporting rock particles—flaked from meter-sized boulders ejected from Earth into outer space by a large asteroid impact on the planet's landscape—could travel to another stellar system within the Milky Way Galaxy. He alleged that such life-forms, unmodified by Homo sapiens, could become a colony if falling onto another Earth-like celestial body (Napier, 2004). Summarizing, Dr. Napier asserts that life's protective mini-spacecraft—rock particles derived from fragmented country-rock—are actually formed by the same outer space weather that is the threat to life's interstellar journey! Thus, like others, Team Geographos assumes interplanetary Lithopanspermia is an operative Astrobiology factor (Cockell et al., 2007; Horneck et al., 2008; Mendonca, 2014; Leighs et al., 2014).

However, there are additional life-flight initiation mechanisms for ejecting life-carrying friable boulders to be flung away, outwards from our bioshell, and into interplanetary space, and possibly with a constant boost from our Sun's unceasing solar wind even into interstellar space. These mechanisms may, or may not, be adequate to induce Earthly life's interstellar migration. Non-photosynthetic micro-organisms living inside rocks have the best chance to survive ejection from Earth's surface as well as a friction-heated fall onto another planet's crust or ocean—a conclusion reported by Dr. Dina Pasini at the European Planetary Science Congress, 8-13 September 2013 held in London, United Kingdom. Team Geographos offers a closing, almost science-fiction, somewhat wild Macro-Imagineering concept: what if a very large asteroid hit the North Atlantic Ocean (Ward and Asphaug, 2003) near the isolated Svalbard Global Seed Vault (78^0 North latitude by 15^0 East longitude), **Figure 6**, a collection of 4.5 million varieties of crops deposited in the seed-bank sent from every ecosystem-nation extant in the world today and, consequently, ejected from the Earth that island's intact concrete and steel crop-seed vault to outer space?

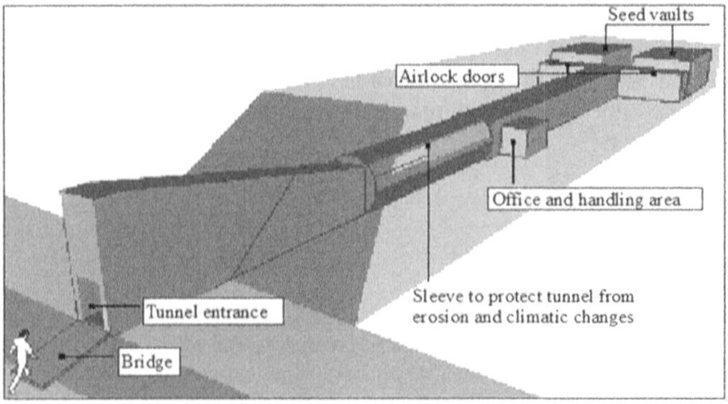

Figure 6. Physical layout of the Svalbard Global Seed Vault located within the Earth's northern Polar Zone. Notice the necessity for protective air-locks to forestall outside contamination as well as for interior air temperature stabilization functions. (Google Image.)

Gun-propelled Anthropic Space Probes

Sir Isaac Newton (1643-1727) cogitated on the physics of powerful imaginary cannon, sited atop a very high mountain peak that projected above our Earth's atmosphere, as an effective means for the emplacement of a planet-orbiting satellite in ***Principia Mathematica*** (Johnson, 1991). Possibly the first purely spallation anthropogenic object ejection—the object was at least partially melted, bent, possibly shaped into a cone and fractured Team Geographos supposes—to escape Earth's gravity, according to Dr. Robert Rex Brownlee (born 1924) on 27 August, 1957, a few months before the internationally-recognized official start of humankind's Space Age with a known orbiting reporting object, **Sputnik-1**! [On 22 March 1998 the Asteroid 15970 "robertbrownlee", provisionally designated 1998FA$_9$, was discovered and named by the Spacewatch astronomical survey group situated at the Kitt Peak National Observatory in the State of Arizona to honor Dr. Brownlee who spent most of his career working at the famed Los Alamos National Laboratory located in the State of New Mexico. **Figure 7a and 7b**.]

Figure 7a. Images personally provided to RBC by Dr. Robert Rex Brownlee, February 2017. Above: The photograph on the left-side was taken about 1969 whilst the drawing was chalked about 1982. **7b.** Below: actual useable U.S.A. postage stamps used by Dr. Brownlee to playfully threaten to constantly "surveil" his grand-children's behavior, from about 1985 to 1988!

By post-event calculation, a circular, iron vertical mine shaft cap sealing a vacuum pipe, approximately 31.2 milliseconds after nuclear device's detonation, was blown—possibly edge-on like southern Africa's anciently incoming Hoba meteorite (Beech, 2013)—into near-Earth outer space by the tamped Pascal-B underground nuclear device test explosion in Area U3d of the U.S.A.'s continental Nevada Test Site. By photographic measurement, not back-of-the-envelope or laboratory calculation, the cap achieved five times Earth's gravitational escape velocity, a hypervelocity of about 56 kilometers per second. How long it remained in outer space near the Earth, or if it truly achieved low eccentricity Earth-orbit, is historically unverified. And, did it carry life-forms into outer space (Huang et al., 2016)? The nuclear test explosive was carefully emplaced in the open-bored and concrete-braced excavation 153 meters below the vacuum shaft-sealing 10.16 centimeter-thick iron cap. The subterranean impulsive nuclear explosion event-process had a shock yield of approximately the equivalent of 272 metric tonnes of TNT. (The conventional explosive TNT is the standard of measurement and comparison for both nuclear and non-nuclear explosions. For another, similar, explosive technique for expelling materials, SEE: "Nuclear-sized explosions without radiation", U.S.A. Patent 4,873,928 awarded on 17 October 1989 to Frank E. Lowther.) By the early 1960s, Gerald Vincent Bull (1928-1990) had developed and achieved the safe missile gun-launch method as an efficient and cost-effective means to regularly deliver small reporting objects to outer space (Gilreath et al., 1999). During the final decade of the 20th Century, John Hunter proposed the construction of the "Jules Verne Launcher", a gun capable of shooting rocket-assisted fired shells in low-Earth orbit; Dr. John Hunter's company, the Jules Verne Launcher Company based in San Diego, State of California, U.S.A., existed as a functional corporate entity from 1996 until about 2012 before becoming dormant or defunct.

A gun launcher devised by Arthur J. Gram Jr. and Charles S. Smith for the American firm Babcock & Wilcox Company—in 2017 celebrating its 150th anniversary after its first establishment during 1867—was assigned U.S.A. Patent 3,131,597 on 5 May 1964 for "Method of and Apparatus for Launching Missiles". The Gram-Smith invention would cause heated freshwater to vaporize, becoming super-pressured steam. Some of the steam then was to react with natural gas to form hydrogen, a part of which was to be piped directly to a high-pressure steam accumulator. When the vertical gun-launcher—installed inside an appropriately-sited stable mountain—was scheduled for discharge, it was expected that steam would compress the hydrogen gas, which would then act to propel the gas expansion impulse-driven projectile upwards, to outer space. Gram and Smith calculated that a 5.4 meter-diameter missile could be launched every three days from a gun barrel with a length of 3,000 meters. Their missile, never exceeding a launch acceleration of 100 gravities, could attain an altitude of 280 kilometers in only 4.5 minutes! Unfortunately, their invention never materialized during the 20th Century, unrealized because the post-**Sputnik-1** technology of rocketry and missile guidance systems advanced so rapidly, overtaking and eclipsing what then seemed like a cruder, even unglamorous Newtonian cannon technology. Gram-Smith foresaw missile launch side-effects occurring within the Earth's bioshell, such as ignition of the suddenly expelled anthropogenic hydrogen gas on contact with the natural air above the launch tube site and subsequent freshwater rain downpours onto the adjacent surrounding countryside! A region of puddled mud landscape generated by benign liquid fallout! In other words, localized torrential rainfall, and the resulting mire on the ground, could make any landscape activities, whether animal, plant or human, messy and costly! Extra-launch site environmental effects, depending on the region in which their first built gun-launcher were located, could also be a significant positive employment factor. Team Geographos lauds this humanistic aspect of the Gramm-Smith mega-project but does not wish its realization in future to cause even a miniature version of the Siberian gas venting that caused a crisis environmental event-process at the end-Permian (Svensen et al., 2009).

"Verneshots" and Our Remarkable "Lava Lite" Whole Earth

Household and night-club lamps called "Lava Lite", first successfully marketed worldwide by the distinctly quirky nudist Edward Craven Walker (1918-2000), remain the popular fixture symbol of the Psychedelic 1960s era of European, North and South American and Asian popular culture. Each lamp is a simple contraption involving an incandescent light-bulb (at its base), a heat transfer coil, a tall, clear glass bottle, and chlorinated paraffin. The paraffin forms its colorful, bobbing blobs, which rise and descend in a water and poly-ethylene glycol background fluid (McClure, 1997). Often, the Lava Lite devices are used in the classroom to convey the Earth-mantle plumes geological concept that may elucidate the chief cause of our planetary crust's tectonic plate spreading and even subducting boundaries. As the internationally-known Serbian astronomer Dr. Milan Milivoje Cirkovic (born 1971) and the American anthropo-geomorphologist R.B. Cathcart (born 1943) have asserted, a really poorly-planned, crust-penetrating "Mission to Earth's Core" macro-project—perhaps an instance of "Predatory Geo-Engineering" (Suarez and van Aalst, 2017)—could actually instigate a large geologic disruption affecting Earthbound humanity adversely (Cirkovic and Cathcart, 2004). It seems reasonable to conclude for now that the sub-aerial Earth-crust and mantle may be, collectively, usefully apprehended as global Nature's "Lava Lite"-like material masses, churning and surging with various fluids and gases that can rapidly affect our Earth's air by sudden massive gaseous outbursts or unfound major leakages! It is even possible that our Earth's global sea-level could be rapidly reduced by undersea methane melting (Bratton,

1999). ["Bubble" usually meant something fragile, unsubstantial, empty or worthless but, nowadays, ordinary persons are urged to be engaged with altering the planet's "Carbon Bubble", a term that first appeared during 2009 to label monetary speculation in government-imposed carbon credits related to fossil fuel consumption.]

Earthly Nature's expelled carbon dioxide gas—the "...largest natural emission of low temperature CO_2 rich gases, from non-volcanic environment, ever measured in the Earth" (Chiodini et al., 2010, page L11303) has been watched for years by attending geoscientists at Mefite d'Ansanto in northern Italy—and methane gas eruptions are well documented (Editor, 2013). However, previous gas eruptions during Earth's earlier Geologic Time far surpassed modern event-processes, apparently, by very deep Earth gas surface breakouts (Ryskin, 2003; Ogden and Sleep, 2012), which were ten times the air's radiation budget imposed by today's carbon dioxide gas content! According to Dr. Jason Phipps Morgan, **Figure 8**, and his colleagues Timothy J. Reston and C.R. Ranero, "Verneshots" are "...catastrophic carbon and sulfur gas-driven craton-lithospheric explosions fracturing a thick lithospheric column, even perhaps capable of shooting large terrestrial mass jets of gas, rocks, soils and dust particles into sub-orbital trajectories in a natural analogue to the explosive transport mechanism first discussed by Jules Verne" (Morgan et al., 2004, page 271). In other words, the energy released from a "Verneshot" may equal that of a large asteroid impact and, as well, that the ballistic transport of an ejected mass could cause a crater elsewhere within the Earth-bioshell, even possibly misinterpreted by geoscientists as a bolide impact! It seems logical to Team Geographos that it is appropriate to extend this "Verneshot" viewpoint (Ravilious, 2004) into the realm of interplanetary Lithopanspermia, at least! And, not to be summarily dismissed as a possibility: "Verneshot"-created planetary rings for Earth that cause shading and, therefore, global climatological changes through sunlight attenuation, a hypothesis not yet properly assessed in an ancient Geologic Time framework, only in terms of modern-day regional and planetary Geo-Engineering (Pearson et al., 2006). The Tropic Zone could have been shaded by any Nature-made ring of small particles located at 1.2-1.6 Earth radii, cooling the Earth's air by shading from sunlight, by reflection lighting the nighttime sky more than a full-Moon, and even altering the intensity of our planet's radiation belts which might affect biotic mutation within the planetary bioshell! Those particles that re-entered Earth's air, like ordinary cosmic dust particles with water-rich clay minerals, will form a melted rock foam bubbles—technically called "vesicular parachutes:, allowing arrival at the Earth's surface (Genge, 2017). Mars might someday be subjected to unnatural "Verneshots" by Homo sapiens' use of nuclear explosives in that planet's future terraformation—that is, dust and small pebbles might escape Mars's gravity and ultimately land on Earth (Mole, 1995). [Stephen Trimberger enjoyed the status conferred by a

U.S.A. Patent Application Publication, US 2011/0005422 A1, on 13 January 2011 for his "Method and Apparatus for Cooling a Planet". Like Mole's concept, dust-raising can warm a planet too. The industrialist entrepreneur Elon Musk, on 9 September 2015, as a guest of ***The Late Show with Stephen Colbert*** television show, offered to warm Mars quickly: "Drop thermonuclear weapons over the poles."] Will Mars eventually be given an artificial ring of orbiting matter that is not merely junked anthropogenic satellites? Measureable radioactivity could then be used to accurately date the onset during Mars' Geologic Time of its quasi ceno-Anthropocene!

Figure 8. Dr. Jason Phipps Morgan, circa 2017. The hypothetical or theoretical, depending on the reader's convictions, "Verneshot" event-process idea must still be vetted even more thoroughly! (Google Images.)

It is most unlikely that Homo sapiens will ever wish, want, or need to totally control the Earth's hot material innards below the planet's crust stratum. But, eventually, it is probable that humans or their robotic descendants will pilot our Solar System, including its intact "Spaceship Earth" within the Milky Way Galaxy (Badescu and Cathcart, 2006). [If we shift the position in the Milky Way Galaxy of our Solar System, we will take our space weather, generated by our Sun, with us: "space weather"—that is, the unceasing solar wind, was discovered by the American scientist Dr. Eugen Parker (Obridko and Vaisberg, 2017).] Even before the extra-Solar System ceno-Anthropocene of Homo astronauticus, such stellar movement of a large object or collection of objects including a shiny star, could be readily spotted by Aliens (Carrigan, 2010; Jones, 2015; Villarroel et al., 2016). ***The Light Year***, a story penned by Italo Calvino (1923-1985), amusingly postulated a cosmic paranoia ensuing after a sign is seen in a far-off galaxy which reads: "I saw you". So, for as long "we", or some embodiment of "we" in physical form allowing for collective humanity to inhabit the Earth bioshell, "we" will have to closely and constantly monitor its subterranean natural and anthropogenic workings in real-time. In the future, we may also have to accommodate ourselves to a sudden-onset, possibly unpredicted, major deep Earth gas eruption of the type hypothesized by Jason Phipps Morgan and others. Obviously, two event-processes will affect the mechanics of anticipated "Verneshots": [I] anthropogenic earthquakes and [II] stress-relief tremors owing to heating caused by technology-induced movements of other planets in this stellar system, such as imposed acceleration, and even of the Earth in the absence, say, of its Moon. Those interested in making Macro-Engineering into a civilization-sustaining profession still have a very long-term travail, as well as a length educational path to trod, to have their knowledge become sufficiently useful in these enormous shifts of mass spatially—that is, globally, perhaps multi-globally, accepted and revered technical and sociological expertise (Grundmann, 2017)! It is very interesting to Team Geographos that the Lava Lite is a model for super-computations of the sort actually required to assess the past and mitigate the future postulated "Verneshots" (Pickover, 1993; Castelvecchi, 2017). The word "model" is always both noun (the digital model) and verb (physical model). If human civilization can build a representative working physical model, it is likely that we can construct the real thing at human scale, but tweakable super-computer digital models can still leave Macro-Imagineering and Macro-Engineering mega-project planners puzzled!

Possibly Team Geographos' idea for a revised Oksiuta-style ***Galactic Space Garden***, briefly described in Chapter One, might be stimulated by the oncoming Asteroid 99942 Apophis. The asteroid has a slight possibility for destructively impacting Earth's bioshell during AD 2029 (Shuvalov et al., 2017). Art's best unintended symbol for this possible near-term future event-process is taken by Team Geographos to be the really big granite rock, **Figure 9**, menacingly balanced several meters directly above human walkers strolling underneath it through a body movement-restrictive concrete slot, visitors to the Los Angeles County Museum of Art situated next to the **La Brea Tar Pits**, where mastodons and sabre-tooth cats (Brown et al., 2017) once perished by asphyxiation caused by slow total immersion in pooled gooey petroleum residue, well previous to 20th and 21st Century ceno-Anthropocene industrial smog in which Los Angeles Basin people are still immersed and unhealthily inhaling during 2018 AD!

Figure 9. ***Levitated Mass*** (2012), selected, tidied and shifted by Michael Heizer (Chianese, 2013; Goodyear, 2016). Weight of metal cleat-suspended boulder = 309 tonnes. The State of California is known for its strong earthquakes! "The sheer size of Heizer's intervention and the infrastructural interruptions it required [to be emplaced for safe public display] led to a degree of public attention rare for other works of art" (Kett, 2015, page 119). Team Geographos imagines this fixed boulder to be an infant-sized representation of Mars's moon Phobos. (Google Image.)

References Cited

Afshordi, N. et al. (27 January 2017) From Planck Data to Planck Era: Observational Tests of Holographic Cosmology. *Physical Review Letters* 118: 041301.

Almar, I. and Shuch, H.P. (2007) The San Marino Scale: A new analytical tool for assessing transmission risk. *Acta Astronautica* 60: 57-59.

Anderson, D. et al. (2016) The least uncomfortable journey from a to b. *American Journal of Physics* 84: 690-695.

Anker, P. (2005) The closed world of ecological architecture. *The Journal of Architecture* 10: 527-552.

Anon. (2006) Tomas Saraceno: Tanya Bonaakdar Gallery. *Artform International* 45: 263.

Armstrong, R. (Ed.) (2017) *Star Ark: A Living, Self-Sustaining Spaceship.* The Netherlands: Springer. 492 pages.

Badescu, V. and Cathcart, R.B. (2006) Use of Class A and Class C Stellar Engines to Control Sun Movement in the Galaxy. *Acta Astronautica* 58: 119-129.

Baker, D.N. and Lanzerotti, L.J. (March 2016) Resource Letter SW1: Space Weather. *American Journal of Physics* 84: 166.

Bartlett, M.D. et al. (13 February 2017) High thermal conductivity in soft elastomers with elongated liquid metal inclusions. *Proceedings of the National Academy of Sciences*. DOI: 10.1073/pnas.1616377114.

Beech, M. (November 2013) Towards an Understanding of the Fall Circumstances of the Hoba Meteorite. *Earth, Moon and Planets* 111: 15-30.

Borisov, G. et al. (2017) The olivine-dominated composition of the Eureka family of Mars Trojan asteroids. *Monthly Notices of the Royal Astronomical Society* 466: 489-495.

Bratton, J.F. (1999) Clathrate eustasy: Methane hydrate melting as a mechanism for geologically rapid sea-level fall. *Geology* 27: 915-918.

Brown, C. et al. (2017) Skeletal trauma reflects hunting behavior in extinct sabre-tooth cats and dire wolves. *Nature Ecology & Evolution* 1: 0131.

Burdorf, L.D.W. et al. (2017) Long-distance electron transport occurs globally in marine sediments. *Biogeosciences* 14: 683-701.

Cadogan, D.P. (September-October 2015) The Past and Future Space Suit. *American Scintist* 103: 338-347.

Calafat, A.M. et al. (January 2008) Exposures of the US Population to bisphenol A and 4-tertiary-octyiphenol: 2003-2004. *Environmental Health Perspectives* 116: 39-44.

Carrigan, R.A. (March 2010) Starry Messages: Searching for Signatures of Interstellar Archaeology. *Journal of the British Interplanetary Society* 63: 90-103.

Castelvecchi, D. (2 February 2017) Astronomers explore uses for AI-generated images. *Nature* 542: 16-17.

Cathcart, R.B. (July 1983) A Megastructural End to Geologic Time. *Journal of the British Interplanetary Society* 36: 291-297.

Cathcart, R.B. and Cirkovic, M. (2006) Extreme Climate Control Membrane Structures: Nth Degree Macro-Engineering. Pages 151-174 ***IN*** Badescu, V., Cathcart, R.B. and Schuiling, R.D. (Eds). *Macro-Engineering: A Challenge for The Future.* The Netherlands: Springer. 316 pages.

Cathcart, R.B. (2011) Anthropic Rock: A Brief History. *History of Geo- and Space Sciences* 2: 57-74.

Cathcart, R.B. (2013) Shaped Metal Earth-Delivery Systems. Chapter 22, pages 507-537 ***IN*** Badescu, V. (Ed.) *Asteroids: Prospective Energy and Material Resources.* The Netherlands: Springer. 689 pages.

Chechetka, S.A. et al. (9 February 2017) Materially Engineered Artificial Pollinators. *Chem* 2: 224-239.

Chianese, R.L. (July-August 2013) Levitated Mass. *American Scientist* 101: 268-269.

Chiodini, G. et al. (2010) Non-volcanic CO_2 Earth degassing: Case of Mefite d'Ansanto (southern Apennines), Italy. *Geophysical Research Letters* 37: L11303.

Cirkovic, M.M. and Cathcart, R.B. (May/June 2004) Geo-Engineering Gone Awry: A New Partial Solution of Fermi's Paradox. *Journal of the British Interplanetary Society* 209-215.

Cockell, C.S. and Lee, M. (2002) Interstellar Predation. *Journal of the British Interplanetary Society* 55: 8-20.

Cockell, C.S. et al. (2007) The Interplanetary Exchange of Photosynthesis. *Origins of Life and Evolution of Biospheres* 38: 87-104.

Cockell, C.S. (July 2010) Geomicrobiology beyond Earth: microbe-mineral interactions in space exploration and settlement. *Trends in Microbiology* 18: 308-314.

Cohen, C. et al. (7 March 2017) On the shape of giant soap bubbles. *Proceedings of the National Academy of Sciences* 114: 2515-2519.

Crossley, K. et al. (1979) An Outbreak of Infections Caused by Strains of Staphylococcus aureus Resistant to Methicillin and Aminoglycosides. II. Epidemiologic Studies. *The Journal of Infectious Diseases* 139: 280-287.

Cullis, P. et al. (February 2017) Pop Goes the Balloon! What Happens when a Weather Balloon Reaches 30,000 m asl? *Bulletin of the American Meteorological Society* 98: 216-217.

DeZern, A.E. et al. (March 2017) Alternative Donor Transplantation with High-Dose Post-Transplantation Cyclophosphamide for Refractory Severe Aplastic Anemia. *Biology of Blood and Marrow Transplantation* 23: 498-504.

Diels, L. (1914) Die Algen-Vegetation der Sudtiroler Dolomitriffe. *Berichte der Deutschen Botanischen Gesellschaft* 32: 502-526.

Dobretsov, N.L. (January 2017) The evolution of planets: Venus as the Earth's Probable future. *Russian Geology and Geophysics* 58: 1-11.

Durairajan, R. et al. (October 2015) InterTubes: A Study of the US Long-Haul Fiber-optic Infrastructure. *ACM SIGCOMM Computer Communication Review-SIGCOMM'15* 45: 565-578.

Editor (August 2013) Quake after-effects. *Nature Geoscience* 6: 585.

Erlich, Y. and Zielinski, D. (3 March 2017) DNA Fountain enables a robust and efficient storage architecture. *Science* 355: 950-954.

Extance, A. (1 September 2016) Digital DNA: Could the Molecule Known for Storing Genetic Information also Store the World's Data? *Nature* 537: 22-24.

Falkowski, P.G. (2015) *Life's Engines: How Microbes Made Earth Habitable.* Princeton NJ: Princeton University Press. 224 pages.

Field, F. et al. (2014) Space-Time and Architecture. *Journal of the British Interplanetary Society* 67: 322-331.

Franks, S.J. et al. (2008) The Resurrection Initiative: Storing Ancestral Genotypes to Capture Evolution in Action. *BioScience* 58: 870-873.

Frazier, J.G. (1997) Sustainable Development: Modern Elixir or Sack Dress? *Environmental Conservation* 24: 182-193.

Fukumori, I. (January 2017) Kamifusen, the self-inflating Japanese paper balloon. *Physics Today* 70: 78-79.

Gefter, A. (26 May 2007). The cosmos as canvas. *New Scientist* 194: 52-53.

Genge, M.J. (2017) An increased abundance of micrometeorites on Earth owning to vesicular parachutes. *Geophysical Research Letters* 44: 1679-1686.

Gertz, J. (29 January 2017) Post-Detection SETI Protocols & METI: The Time Has Come To Regulate Them Both. Posting at website ARXIV.ORG 1701.08422.

Gibbs, W.W. (2004) Synthetic Life. *Scientific American* 290: 74-81.

Gillings, M.R. and Paulsen, I.T. (2014) Microbiology of the Anthropocene. *Anthropocene* 5: 1-8.

Gilreath, H.E. et al. (1999) Gun-Launched Satellites. *Johns Hopkins Aplf, Technical Digest* 20: 305-319.

Goddard, E.C and Pendray, G.E. (Eds.) (1970) The Last Migration. *The Papers of Robert H. Goddard.* Volume III, 1938-1945, pages 1611-1612. New York: McGraw-Hill.

Gold, Thomas [1920-2004] (May 1960) Cosmic Garbage. *Air Force And Space Digest*, page 65.

Goldstein, A.H. and Galbally, I.E. (2007) Known and Unexplored Organic Constituents in the Earth's Atmosphere. *Environmental Science & Technology* 4j0: 1514-1521.

Goodyear, D. (29 August 2016) Profiles: THE EARTH MOVER. *The New Yorker*. Page 60-71.

Gracie, A. (2016) Art, Space and Hyperreality: An Artistic Exploration of Artificiality, Meaning and Boundaries within Astrobiological Practice. *Leonardo* 49: 6-33.

Grundmann, R. (2017) The Problem of Expertise in Knowledge Societies. *Minerva* 55: 25-48.

Haber, W. (2007) Energy, food, and land—the ecological traps of humankind. *Environmental Science and Pollution Research* 14: 359-365.

Hansson, A. (1996) Towards Living Spacecrafts. *Journal of the British Interplanetary Society* 49: 387-390.

Hart, R.M. (2013) *Progress in Environment Suits.* NY: CordaNobello. 242 pages.

Homann, M. et al. (December 2015) Evidence for cavity-dwelling microbial life in 3.22 Ga tidal deposits. *Geology* 44: 51-54.

Horneck, G. et al. (2008) Microbial Rock Inhabitants Survive Hypervelocity Impacts on Mars-Like Host Planets: First Phase of Lithopanspermia Experimentally Tested. *Astrobiology* 8: 17-44.

Huang, Q. et al. (2016) A molecular perspective on the limits of life: Enzymes under pressure. *Condensed Matter Physics* 19: 22801-1.

Hutchinson, C.A. et al. (25 March 2016) Design and synthesis of a minimal bacterial genome. *Science* 351: 1414.

Johnson, W. (1991) Some Monster Guns and Unconventional Variations. *International Journal of Impact Engineering* 11: 401-439.

Jones, M. (November-December 2015) Reconsidering Macro-Artefacts in SETI Searches. *Acta Astronautica* 116: 161-165.

Jutte, D. (2015) *The Strait Gate: Thresholds and Power in Western History*. London: Reaktion Books. 384 pages.

Kett, R.J. (Spring 2015) Monumentality as Method: Archaeology and Land Art in the Cold War. *Representations* 130: 119-151.

Kirkby, M.J. (2016) Water in the critical zone: soil, water and life profile to planet. *SOIL* 2: 631-645.

Koppelmans, V. et al. (19 December 2016) Brain structural plasticity with spaceflight. *Npj Microgravity* 2: 1. DOI: 10.1038/s41526-016-0001-9.

Kriegler, E. et al. (2013) Is atmospheric carbon dioxide removal a game changer for climate change mitigation? *Climatic Change* 118: 45-57

Kwan, J. et al. (2010) The Adventures of the Rocketeer: Accelerated Motion Under the Influence of Expanding Space. *Publications of the Astronomical Society of Australia* 27: 15-22.

Lan, L. et al. (23 November 2015) Debris Engine: A Potential Thruster for Space Debris Removal. Posting at website ARXIV.ORG: 1511.07246.

Landenmark, H.K.E. et al. (2015) An Estimate of the Total DNA in the Biosphere. *PLOS Biology* 13: e1002168.

Leighs, J.A. et al. (2014) The bactericidal effect of shock waves. *Journal of Physics: Conference Series* 500: 182026.

Liebenberg, L.W. (2006) Persistence hunting by modern hunter-gatherers. *Current Anthropology* 47: 1017-1025.

Longo, G. (February 2017) The web and its sorceries. *AI & Society* 32: 135-136.

Low, S. (23 December 2016) The future of climate engineering. *Earth's Future* 4: 1-4. DOI: 10.1002/2016EF000442.

MacMartin, D.G. et al. (2016) Geoengineering with stratospheric aerosols: What do we not know after a decade of research? *Earth's Future* 4: 1-6.

Mautner, M.N. (2005) Life in the Cosmological Future: Resources, Biomass and Populations. *Journal of the British Interplanetary Society* 58: 167-180.

McClure, M. (1997) Lava Lite Lamps: A Chemical Juggling Act. *Chem Matters* 15: 4-7.

Mendonca, M. de Souza (1 May 2014) Spatial ecology goes to space: Metabiospheres. *Icarus* 233: 348-351.

Millard, D. (2017) *Satellite: Innovation in Orbit.* London: Reaktion Books. 206 pages.

Miyake, F. et al. (31 January 2017) Large ^{14}C excursion in 5480 BC indicates an abnormal sun in the mid-Holocene. *Proceedings of the National Academy of Sciences* 114: 881-884.

Mole, R.A. (1995) Terraforming Mars with Four War-Surplus Bombs. *Journal of the British Interplanetary Society* 48: 321.

Morgan, J.P. et al. (15 January 2004) Contemporaneous mass extinctions, continental flood basalts, and 'impact signals': Are mantle plume-induced lithospheric gas explosions the causal link? *Earth and Planetary Science Letters* 217: 263-284.

Napier, W.M. (2004) A Mechanism for Interstellar Panspermia. *Monthly Notices of the Royal Astronomical Society* 348: 4651.

Neudeck, P.G. et al. (2016) Prolonged silicon carbide integrated circuit operation in Venus surface atmospheric conditions. *AIP ADVANCES* 6: 125119.

Obridko, V.N. and Vaisberg, O.L. (March 2017) On the history of the solar wind discovery. *Solar System Research* 51: 165-169.

Ocampo, R.P. and Klaus, D.M. (2016) Comparing the Relative Risk of Spaceflight to Terrestrial Modes of Transportation and Adventure Sport Activities. *New Space* 4: 190-197.

Ogden, D.E. and Sleep, N.H. (3 January 2012) Explosive eruption of coal and basalt and the end-Permian mass extinction. *Proceedings of the National Academy of Sciences* 109: 59-62.

Oksiuta, Z. (2007) New Biological Habitats in the Biosphere and in Space. *Leonardo* 40: 122-123.

Owens, M.J. et al. (2017) Global solar wind variations over the last four centuries. *Scientific Reports* 7: 41548.

Parker, P. (1999) Millennium Rock. *Spaceflight* 41: 262.

Pearson, J. et al., (January 2006) Earth rings for planetary environment control. *Acta Astronautiaca* 58: 44-57.

Philippe, J-M. (1999) The KEO Satellite. *IAF-99-P.3.10.*

Pickover, C.A. (1993) Lava Lamps of the 21st Century. *The Visual Computer* 10: 173-177.

Prairie, Y.T. and Duarte, C.M. (2007). Direct and Indirect metabolic CO_2 release by humanity. *Biogeosciences* 4: 215-217.

Pulido, A. et al. (22 March 2017) Functional materials discovery using energy-structure-function maps. *Nature.* DOI: 10.1038/nature21419.

Raitt, D. (2007) ESA and the Arts: A Programme in the Making. *Acta Astronautica* 60: 48-54.

Ravilious, K. (8 May 2004) Four days that shook the world. *New Scientist* 182: 32-35.

Reasoner, D.I. Chemical-release mission of CRRES [Combined Release and Radiation Effects Satellite]. *Journal of Spacecraft and Rockets* 29: 580-584.

Ryskin, G. (September 2003) Methane-driven oceanic eruptions and mass extinctions. *Geology* 31: 741-744.

Sagullenko, E. et al. (1 February 2017) Nuclear Pore-like Structures in Compartmentalized Bacterium. *PLOSOne* 12: e0169432.

Sayre, N.F. (2008) The Genesis, History, and Limits of Carrying Capacity. *Annals of the Association of American Geographers* 98: 120-134.

Scheil, B. (July-August, 2008) The Memory of an Elephant. *Architectural Design* 78: 44-49.

Schrijver, C.J. (2015) Socio-Economic Hazards and Impacts of Space Weather: The Important Range Between Mild and Extreme. *Space Weather* 13: 6-9.

Sheppard, D.J. (1979) Concrete Space Colonies. *Spaceflight* 21: 3-8.

Shin, J.K. (2016) Application of Cellular Automata for a Generative Art System. *Leonardo* 49: 396-430.

Shuvalov, V.V. et al. (January 2017) Asteroid Apophis: Evaluating the impact hazards of such bodies. *Solar System Research* 51: 44-58.

Sole, R.V. et al. (2015) Synthetic circuit designs for Earth terraformation. *Biology Direct* 10: 37.

Steiner, H. (2005) The Forces of Matter. *The Journal of Architecture* 10: 91-109.

Suarez, P. and van Aalst, M.K. (2016) Geoengineering: A Humanitarian concern. *Earth's Future* 5: 1-13. DOI: 10.1002/2016EF000464.

Sukumaran, J. and Knowles, L.L. (January 2017) Multispecies coalescent delimits structure, not species. *Proceedings of the National Academy of Sciences.* DOI: 10.1073/pnas.1607921114.

Svensen, H. et al. (30 January 2009) Siberian gas venting and the end-Permian environmental crisis. *Earth and Planetary Science Letters* 277: 490-500.

Teclaff, L.A. (1994) Beyond Restoration—The Case of Ecocide. *Natural Resources Journal* 34: 933-956.

Vakoch, D.A. (2016) In defense of METI. *Nature Physics* 12: 890.

Van Lemmen, H. (2006) *Coade Stone.* UK: Shire Publications Ltd. 48 pages.

Vanlint, D. et al. (2011) mBio 2: e00130-10.

Verseux, C.N. et al. (2016) Synthetic Biology for Space Explorations: Promises and Societal Implications. *Ambivalences of Creating Life* 45: 73-100.

Villarroel, B. et al. (29 June 2016) Our Sky now and then—searches for lost stars and impossible effects as probes of advanced extra-terrestrial civilizations. ARXIV. ORG posting 1606.08992.

Ward, S.N. and Asphaug, E. (June 2003) Asteroid impact tsunami of 2880 March 16. *Geological Journal International* 153: F6-F10.

Wolfe, A.J. (2002) Germs in Space: Joshua Lederberg [1925-2008], Exobiology, and the Public Imagination, 1958-1964. *Isis* 93: 183-205.

Wright, J.T. (24 April 2017) Prior Indigenous Technological Species. *ArXiv posting 1704.07263v1.*

Yazdi, S.M.H.T. et al. (2015) A Rewritable, Random-Access DNA-Based Storage System. *Scientific Reports* 5: 14138.

Yeomans, J.M. (2017) Nature's Engines: Active Matter. *EPN* 48(2): 21-25.

Yokoo, H. and Oshima, R. (1979) Is bacteriophage X174 DNA a message from an extraterrestrial intelligence? *Icarus* 38: 148-153.

Zalasiewicz, J. et al. (June 2014) Human bioturbation, and the subterranean landscape of the Anthropocene. *Anthropocene* 6: 3-9.

CHAPTER TWO

Areographic Climate Control: Humanity's Mars Colonization Macro-Project

Our Universe's Cosmic Micro-Wave Background, when imaged, might be perceived as a colorful colony of Petri-dish bacteria examined by microscope! The same might be said of Earth-soil particles! Our planet's varied-texture landscapes incorporate a large volume of carbon in inorganic forms in a matrix which absorbs most visible wave-lengths of sunlight, giving many named soil-types their usual brown-hued coloration (Allison, 2006). For thousands of years people have tended, piled, and dug Earth's soils for agricultural as well as destructive and constructive uses and relied upon the Earth epidermis' long-term inorganic and organic continuity (Richter, 2007). Probably seventy million square kilometers of our planet's ice-free landscape has been directly modified by Homo sapiens actions (Harden, 2014)! Biotic event-processes demonstrably affect landforms at every geographical scale, from tiny creatures inside rocks to larger realms, yet there presently seems to be no single large-scale landform that, uniquely as well as indisputably, indicates the presence of non-Homo sapiens biota (Tarolli and Sofia, 2016)—excepting perhaps beaver mud dams, termite mounds and migrating herd animal landscape erosion tracks. Homo sapiens has bored into the deep-sea floor and built artificial mountain ranges with skyscrapers. Humans are a major soil-forming factor (Yaalon, 2007), even by the erosion over extended periods of time of our tallest buildings: the hastily abandoned town of Pripyat near radioactive Chernobyl (1986) and the disused, concrete-coated 6.47 hectare Hashimi entrance to the undersea coal-mine company-town island near Nagasaki are being converted by Nature into soil; the Mitsubishi Corporation operated its Hashimi facility until its closure and abandonment during 1974. Isolated and forbidding, it was a fictional villain's spooky lair in the 2012 James Bond movie ***Skyfall***.

A planet is an approximately spherical solid object moving in interplanetary or interstellar space that is smaller than a star—a planet has never undergone nuclear fusion during its Geologic Time and it has enough mass so that its masses of various components has sufficient self-gravitation to naturally assume a generally globular shape, usually fine-tuned by rotation. Interestingly, the English noun "homeworld" only just barely antedates AD 1945, and usually it is applied only to Earth. By mid-19th Century, Thomas Dick (1774-1857) postulated a natural Mars, utterly unaltered by Homo sapiens except by naked-eye and telescopic human gaze and all the assumptions generated by viewing, might host a population of 15.5 billion persons, whereas Dr. Paul Birch foresees a 21st Century-transformed Mars—quickly terraformed during an intensive half-century period of applied Macro-Engineering—ultimately becoming an invaded second planetaric Homo sapiens homeland, where special survival-style clothing, of the fashionable inflationary muscles and miniaturized electronics variety (Maziz et al. 2017), would eventually not be required by its settled and established pioneering human inhabitants for living healthily (Birch, 1992). Our planet's still incompletely utilized bioshell may, by AD 2100, house 10 billion persons. Therefore, in Team Geographos' estimation, it is perfectly reasonable to envision a thrilling and inspiring near-term future serious Mars colonization effort as equivalent to the U.S.A.'s 19th Century so-called "Wild West" settlement. In fact, Chapter Two visualizes 2,000,000 happy Homo sapiens Martians, some migrants and some native, living comfortably on Mars before AD 2101, the true calendrical start of the 22nd Century. The term "Areography", a useful pre-Space Age modification of "Geography" was used in English as early as 1868 by astronomer Richard Anthony Proctor (1837-1888), simply substitutes the Greek name for frigid, dry Mars [Ares] for that of Earth [Geo].

The driest part of Chile's central Atacama Desert, nearly devoid of soil in the context of ordinary agricultural assessment, is currently presumed to be the limit for non-dormant sub-aerial life-forms. Laboratory studies proved Earth's methanothermobacter wolfeii could survive in the subterranean groundwater deposits of today's Mars (Sinha et al., 2017). By future human Brobdingnagian macro-projects—made doable by advanced regolith sintering technology—designed to increase the moisture and carbon content of Mars' future manipulated ground surface, that planet may undergo an anthropogenic tinctumutation (from orange to some hue of brown) during the 21st Century that will be visually discernable by interested human naked-eye sky-watchers! For its prosperous Homo sapiens settlers, Mars need never remain a monotonous, flat red-orange colored landscape, as the anagram of "terraforming" suggests, "art [will] reign form"; "reign" indicates a period of human rule [as of a dynasty of monarchs]. Team Geographos is strongly inclined to use a constructed acronym, TAMP, a label derived from the phrase "para-Terraformed Arid Mars Plumbing"—"tamp" means "to brutishly force or pack something" such as Mars' reddish granular ground [dust and pebbles] tighter by tapping with a tool, a tamper, after burying a saline and/or freshwater pipeline. Widespread future ground rearrangements will create, as on our present-day Earth-crust, a topographic signature of humanity's presence. Mars' technical transformation, even possibly its 21st Century revitalization, is a down-to-Earth Macro-Imagineering effort well worth consideration. United Kingdom researcher Dr. Paul Birch's 1992 proposed Mars terraformation mega-project plan seems almost fantastic, but it is a relentlessly logical extension of other suggested Mars vitalization make-over macro-projects. Our Solar System offers, approximately, 1.7 billion square kilometers of solid ground; Earth's surface area is 510 million square kilometers, about 30.06% is ground humans walk on, whilst Mars' area is 145 million square kilometers, or about 8.53% of the Solar System's known ground. It may always be impossible to create Utopia anywhere but any human-established Dystopia on Mars could not exist but momentarily, owing to certain, inevitable erasure by that planet's harsh geophysical realities, specifically in terms measured by the average international human Standard of Living definition (Cockell, 2002; Meyer, 2004). Such failure might be cathartic!

As means for safe transportation of cargo and people between the Earth-Moon system and Mars is technically perfected (Bolonkin, 2005), one of the outcomes will probably be assisted colonization of Mars, owing in part to possible rapid climate regime change within the Earth-bioshell. Capitalizing on human-kind's remarkable history of macro-engineering approaches to overcoming Earth's geophysical challenges, Geo-Engineering techniques ought to be included as viable solutions for reducing the future anticipated economic impacts of Nature's and Homo sapiens' warming of Earth's air (Martin and Pinkyck, 2015, page 2981); indeed, organized real-world outdoor experimentation is already underway with many oceanic, islandic and continental regions to test various symptom alleviation methodologies! In this case, Team Geographos views Geo-Engineering, old [such as the Haber-Bosch process] and modern, applied to the atmosphere as a transformative needed adaptation effort applied as a significantly larger geographical scale and intensity than ever before in the ceno-Anthropocene—Earthly climate "control" can become a truly new multi-global impact event-process! If we assume that Mars' current atmosphere is in a non-anthropogenic "ruin" state, then it seems reasonable to extend our imposed anti-ruination Earth Geo-Engineering techniques to an awaiting Nature-mothballed Mars, its costly enactment done for presently visible compelling reasons. **Figure 1.** By adding Cyanobacteria (class Oxyphotobacteria) (Soo et al., 2017) to split water by photosynthesis, after people have modified Mars, may produce a loveable world.

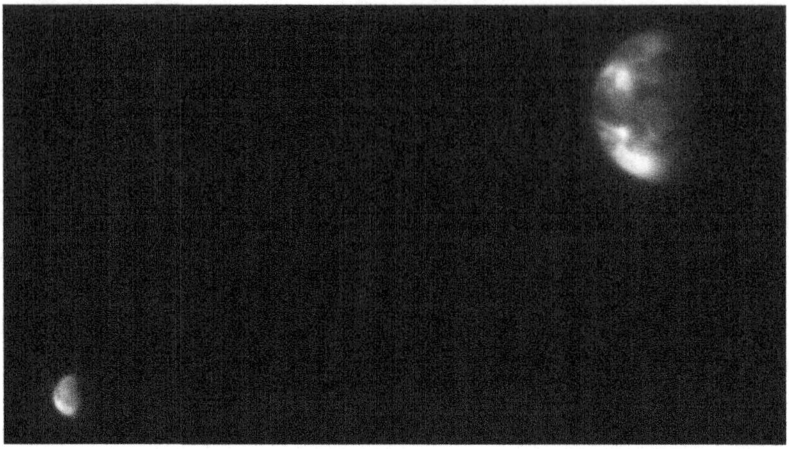

Figure 1. Composite image of the Earth-Moon system taken from the vicinity of Mars on 20 November 2015 by the NASA's High-Resolution Imaging Science Experiment camera. (NASA Image.)

Our first close-up image obtained of Mars' surface was transmitted to Earth from Mars by the U.S.S.R.'s **Mars-3** during 1971; since 1971, more than 9,300 kilograms of anthropogenic materials have since attained final repose, mostly as imposed anthropic junk littering Mars's ground surface (McKinney, 2014). The NASA, ESA, JAXA and other space-faring organizations (Russia, India, China, Israel, Iran et al.) usually give such machine tools a touch of coyly named personification so that any tool's success becomes a victory for aspiring, yet Earth-bonded, grouped Homo sapiens! As of 2017, Mars is still populated only by robots reporting on the Red Planet's extant aerial and sub-aerial geophysical conditions; artificial acceleration of spationaut physiology development and/or amplification of physique and trained skill sets may be prerequisite for truly permanent Homo sapiens colonization of Mars. Humans have been particularly effective in blurring or eliminating biogeographic barriers through long-distance trade, intentionally introducing some species and carrying others as hitchhikers: the resulting translocation of species, now being done almost regularly via spacecraft (both current scrubbed and non-scrubbed space and planetary probes and near-term future spaceships carrying people, each with a peculiar personal microbial biome) will change Mars, for sure. Inside, Mars' colonial housing, the walls will even indicate the presence of humans, a kind of extra-somatic biogeography (Ruiz-Calderon et al., 2016). Cleansing and repair specialists of the dedicated ***Mary Poppins*** character type, as made readably delightful in that 1934 novel by Pamela Lyndon Travers (1899-1996), needs find an honored place within and around such Mars housing and workshop confines! Other than Earth itself, Mars has the most Earth-like landscape within our Solar System. Its days are just 37 minutes longer than Earth's. Average daytime temperatures can be suitable for green plant growth, rising to 14^0 Celsius, but nighttime temperatures can fall as low as minus 63^0 Celsius. Projecting Martian wind-chill in familiar Earth terms, Randall Osczevski (2014, page 534) rightly foresaw that "...intensely bone-chilling image of Mars could become a psychological barrier for potential colonists, as well as public support for such ventures.... Thermometer readings from Mars are highly misleading to terrestrials". Surface insolation is only 43% of Earth's, due to Mars' greater orbital distance from the Sun. Extensive and widespread dense-particle ruddy dust storms reduce sunlight levels by several orders of magnitude for weeks; active Spring and Summer dust devils on Mars—dry convective vertical fast vortices (columnar or cone-shaped) are visible by the concentration of dust lofted—were confirmed more than thirty years ago (Thomas and Gierasch, 1985)—may be predicted to scour all exposed emplaced anthropogenic infrastructure, just as circular winds fictitiously did in ***The Martian*** (2015) Hollywood-made movie (Miller, 2017)! [On Mars of ***The Martian***, the male hero's potato farming—traditionally the onerous food-raising task of a bedraggled Earthly peasant-cultivator—is meant to evoke the paradigmatic Neo-Classical figure of the duteous hypothesized Homo economicus (Strychacz, 2017). Mars won't likely be a Holiday Frontier for the extended tourism industry for some period of time yet.] Supplemental artificial "sun-lighting" of commercial crop production is quite common in Canada and northern Europe. However, the main impediments to plant growth on Mars are low air pressure (7.5 millibar versus Earth's 1,013 at sea-level) and the apparent absence of surface-pooled liquid freshwater. Plants can be cultivated on Mars, but only inside protective structures, well-kept closed environments wherein trace contaminants in the fresh groundwater and ambient air can result from maintaining equipment operation (ozone, hydrocarbons, the human colonizers (methane) or from the grown plants (such as ethylene). Nevertheless, the petroleum-rich ecosystem-state of United Arab Emirates is leading an internationalized R&D effort, ***Project Hope***, meant to place a satellite in orbit of Mars by 2020 AD (El-Showk, 2017); the UAE also hopes to construct a non-formulaic human-inhabited city somewhere on Mars a century hence—that is, by circa 2117 AD.

Bearing in mind that our species will soon be far outside its historic Earth-bioshell natural range, Team Geographos has herein organized a Macro-Imagineering framework that can be used to exemplify potential actions under a suite of feasible Mars future settlement scenarios. In the absence of rainfall—even post-Dr. Paul Birch-style Mars terraforming—Team Geographos suggests artesian basins and the planet's two polar dry-ice/freshwater ice-layered caps (totaling 3,000,000 square kilometers) may serve to provide a two million-person human Mars civilization with a basic, more than 95% pure, freshwater supply melted, tapped and taken from the north polar deposits of Mars (Grima et al., 2009) to a potentially more salubrious equatorial region, possibly with a concentration of high-density human populace frequenting the shoreline of a voluminous anthropogenic lake situated at the very bottom of Valles Marineris where air pressure is highest. In a word used here postpositively—that is, after the noun it modifies—Mars Redivivus, since the Team seeks to partially restore the planet to its supposed ancient state, more beneficial to life-forms, that it once exhibited to the Universe uncounted years ago of its past Geologic Time. To that ambitious goal, Team Geographos offer planetary alteration equipment that is lightweight, compact, power efficient, modular, and capable of performing several varied and vital tasks in support of facility construction and infrastructure in-situ resource utilization. There will be a need for massive dust suppression. Space Exploration Technologies Corporation, a legal creature of industrialist-entrepreneur Elon Musk, has announced plans to have tourists loop around the Moon (Pasztor, 2017) and to be Mars' first spectator tour group! However, lunar industrialization plans by the NASA already include a ground-based site dust suppression mobile sintering robot the presages the orbital devices inherent in Dr. Paul Birch's terraforming scheme. In other words, TAMP—the Terraformed Arid Mars Plumbing mega-project—is a requirement for proper Mars anthropogenic change.

Mars in Our Olden Times: Landscape Hermeneutics

The first printed map of Mars, inscribed by Wilhelm Beer (1797-1850) and Johann H. von Madler (1794-1874), is most notable because of the selection by its makers of a Prime Meridian based on a markedly high-albedo landscape feature near the planet's equator. Their Areographic indicator placement, still in use during the 21st Century by chart and globe makers for practical mapping purposes, is nearly identical with the Mars Prime Meridian, defined ultimately during AD 1873 to pass through Airy-O, a bolide impact crater roughly 500 meters in diameter which lies within a 56 kilometer-diameter meteoric crater named for George Biddell Airy (1801-1892). Internationally reported in print media, observations of Mars from 1892 to 1910, prior to the revelatory non-confirming images transmitted by the NASA Mariner probes during 1965-1972, forged a popular assumption of a hyper-arid Mars landscape fully managed by some kind of intelligent animal—creatures possibly intellectually superior to Homo sapiens (Lane, 2005) using a deliberately dug network of desert canals globally distributing all available liquid freshwater (Lane, 2006). To be visible to Earth's bioshell-confined humans, these canals were then presumed to be very wide and of enormous flowing fluid-transporting capacity; the presumed cropland irrigated in both hemispheres on both sides of these canals was generally thought to accentuate the channels "visible" to Earthly telescopic viewers. Later, it was discovered that some of Mars' "canals" actually did correspond broadly to certain especially pronounced actual landscape features fuzzily imaged by Mariner 9, such as huge-diameter circular craters and wide crust rifts or valleys, as well as some other marked ground albedo features (Gerstbach, 2003). Even during later 20th Century years, approximately 1976 until 1998, humanity was abusively mistreated to the mythology of ***The Face on Mars***, a fascinating pareidolia/mimetolith that helped to publicly impugn by media-published and media-broadcast falsehoods the NASA by surrounding that politicized governmental organization with pestering strange conspiracy theories related to a supposedly unreported Alien civilization on Mars! Since 1997, robots in the form of orbiting satellites, soft-landers and roaming rovers have operated continuously on or around Mars (Vasavada, 2017): there is much of interest, but definitely absent is ***The Face on Mars***! Artists and micro-biologists wish to incorporate human-composed messages in the bacteria Deinococcus radiodurans (Vaidyanathan, 2017), one even meant to invisibly outline the human hand in the innards of Bacillus subtilis (Freymond et al., 1999); were these designed bacteria sent to Mars before humans arrive they would together symbolize that planet's inherent inhabitability as well as a "signed" planet's possession by Homo sapiens! In other words, Micro-Imagineering could temporarily eclipse Macro-Imagineering! A topographical anomaly need not be present—like "canals"—but, clearly, the planet is in possession, at least partially, of an Homo sapiens-altered life-form.

Since the widespread institution of outdoor electric lighting of Earth's major cities since 1882, two million transplanted and native future Homo sapiens Martians should be able to readily observe, using low-magnification amateur telescopes or equally effective binoculars, during clear Mars nights of the early-22nd Century, their artificially illuminated former homeland! Team Geographos believes and thinks such an astronomical spectacle is an advertisement recommending the arduous Homo sapiens duty to complete the terraformation of Mars. For instance, newly arrived settlers might find such visual spectacle helpful in reducing their restrained but ever-present emotions of Mars colonizer's homesickness. The sheer magnitude of unaided eye-detectable unnatural light spillage emitted by Earth's bioshell, over several human generations' time, can serve as a salient, inspirational proxy for humanity's other ongoing economic activities within our Solar System, such as ore mining operations in the Asteroid Belt located slightly beyond the orbit of Mars (Cathcart et al., 2013). Vital supplies necessary to keep people alive and healthy and their supportive machinery running reliably, as well as building materials for these bold new Martians will certainly be extracted from selected suitable asteroids and delivered to Mars' surface by specially designed vehicles [package] wrought from the materials [payload] being shipped to Mars! Since electric outdoor lighting was a novelty during the early years of the 20th Century, and rural farming remained the dominant economic and social endeavor for our globalizing civilization, it is no wonder that astronomers of olden times left generally unconsidered the proposition that any extant "Martians" would cause outdoor light-pollution as urbanized humans do today (Zhou et al., 2015). Operational 21st Century satellites has made it possible for Earth's macro-imagineers and macro-engineers to study Mars in near real-time without benefit of large fixed telescopes—we have exited the domain of Astronomy as well as the realm of data-possessive astronomers! These remotely-sensed and transmitted data are so accurate that our profession anticipates physical models of Mars' fluid catchments [drainage basins] will be built to test how freshwater behaves when something stimulates, disrupts or changes the areic and endoreic periglacial freshwater flows, since even today's marvelous supercomputer numerical simulations result in questionable or inadequate predictions for the macro-projects envisioned to produce a non-facsimile "Earth" of Mars! Retro-diction is a common activity of prefiguring geoscientists and pre-final infrastructure configuration mega-project macro-engineers, used to test and calibrate hypotheses and working theories of past Geologic Time processes and events—indeed, all of the "Palaeo"-prefixed scientific disciplines require its employment! Macro-Imagineering must also use this learning technique to project an integrated industrial mapping of global event-processes (Wolf, 2012; Hadhazy, 2015; Castelvecchi, 2017). Physical models of Mars, inexpensively contrived even in non-pliable or moldable standard LEGO block format, might be cheaply practical for some applications (Wolf, 2014).

Essentially, human geoscientists and macro-project dreamers no longer need to develop and study elaborate theories of interpretation and understanding of the Red Planet's landscape hermeneutics. Instead, the politically-defined settlement regions of a terraformed Mars can conform exactly to Mars' anthropogenic periglacial catchments (Kauffman, 2002; Morris, 2014). "Periglacial" was coined circa AD 1909 by Walery von Lozinski and first presented to participants of the XI International Geological Conference's 1910-1911 an excursion to the Arctic Zone town of Spitzbergen; it applies to landscapes in which frost-related geomorphic event-processes and/or permafrost are dominant or characteristic (French, 2003). There are two main forms of ground-ice: ice that is structure-forming, bonding the enclosed sediments, or large bodies such as Mars' Polar Zone ice-caps that are layered mixtures of frozen carbon dioxide gas and freshwater. High-latitude regolith of Mars has thermal properties conforming theoretically to a high thermal inertia permafrost deposit. Mars' future human settlers are likely to own useful, accurate topographic maps that are much superior to the informative, natural-color commercial airline passenger route maps—the ***Jeppesen Natural-Color Map Series*** produced during the 1950s and 1960s—painted by Hal Shelton (1916-2004). "Because of their detail and realism, [the] NASA used these maps to locate and index photo[graphs] of Earth taken on early space missions" (Patterson and Kelso, 2004, page 28). The impending 21st Century organization of a pioneer human society colonizing Mars catchment-by-catchment successfully melds two fundamental Macro-Engineering landscape concepts into domed and outdoor designer ecosystems: (I) monitored carrying capacity and (II) close conformance to explored and known geological structures as well as known regional variations in regolith composition and regionally compartmentalized groundwater deposits and flows.

During 1666 AD, Giovanni Domenico Cassini (1625-1712) first observed and noted how the white-colored, non-identical in area Polar Zones of Mars waxed and waned seasonally. **Figure 2**. Since 1905, uncountable numbers of Areographic mapping images that show Mars' distinguishable high-latitude Polar Zones have been obtained by geoscientists; yet, it was not until the mid-1960s, after Earth-orbiting satellites acquired images, that mapping of Earth's Polar Zones became viewable by the world-public (Fleming, 2007)! Mars does have a Nature-fostered Hydrologic Cycle, the details of which are still being discovered, evaluated and understood by scientists and macro-engineers. Apparently, it is mostly a subterranean circulatory system, which if contaminated by hardy Earthly microbes might become technically irreversible. Team Geographos rejects this fear as an anti-Technological Fix geo-philosophical lunacy; "reversible exploration" (McKay, 2009; Schon, 2009) is economically impossible and would cramp, possibly even curtail, science's Space Age advances. In terms of Mars, future human settlers wielding high-technology will need to scrabble vigorously, artificially sublimating and efficiently harvesting seasonally receding and advancing Polar Zone ice-caps, digging to extract found frozen groundwater and dust-buried glaciers, and transporting mined, useable freshwater in quantity great areographical distances over the planet's wind-blown dune and rock landscape, generating a unique—first ever—Anthropogenic Hydrologic Cycle in doing so!

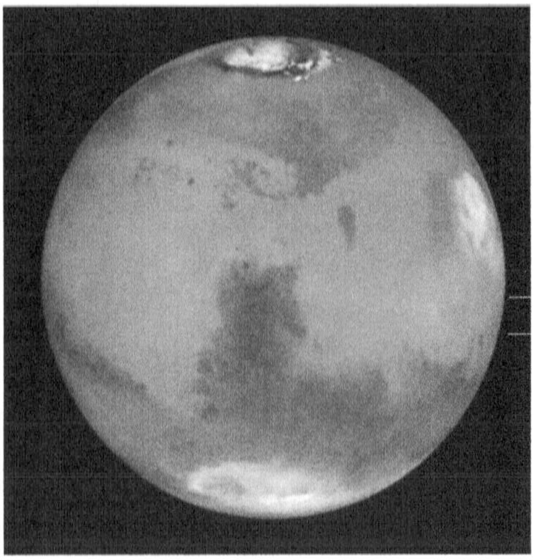

Figure 2. Mars image showing its different area Polar Zone ice-cap regions. (Google Image.)

Pre-conditioning Mars to Receive Homo sapiens in Near-term Future

Self-taught British lay-expert meteorologist Luke Howard (1772-1864) published his practical classification of Earth's aerial cloud forms during 1802-1803. He categorized Nature's clouds as cumulus, stratus, cirrus and nimbus types. Future human Martians might add a fifth type: Anthropogenic clouds. 21st Century architects, Elizabeth Diller and Ricardo Scofidio, constructed their ***Blur Building*** over Lake Neuchatel, near the City of Yverdon-les-Bains, Switzerland as invited exhibitioners at the **Expo.02** held during AD 2002. **Figure 3**. Their nubilous innovative exhibit consisted of vertically-driven lakebed piles—visible, unlike those wood piles that support Venice, Italy—and a horizontal walkway, a constructed pier 100 meters wide by 65 meters deep and 25 meters high, directly above which a fine mist of filtered freshwater, abstracted from Lake Neuchatel, is sprayed into the air from high-pressure nozzles integrated with the cantilevered superstructure. Necessarily, the pier had to conform to the economics of construction imposed by the lake's bathymetry. **Expo.02** visitors, at their hotels, may have been ceremoniously served by a water sommelier—expensive bottled unfiltered freshwater, drawn from domestic and foreign artesian wells and Polar Zone icebergs, but the artificially fog-shrouded ***Blur Building*** was a platform enveloped by a haze of freshwater droplets under autonomous mini-computer monitoring and control, adapted to prevailing outdoor weather conditions to especially please sunny-day strollers. The City of San Francisco, State of California, founded its famed Exploratorium facility in 1969. Since 2013, the Exploratorium has been sited at Pier 15. One of the patrons favorite permanent exhibits it, literally, operates on a timetable was devised by famous Japanese female artist Fujiko Nakaya's delightful scheduled fog-sculpture, ***Fog Bridge*** (Miranda, 2013), which almost replicates ***Blur Building***. Similarly, such unusual architectural experiments will be needed to ensure successful humankind's future long-term settlement of far-distant Mars! Domes composed of Pykrete and hydrogel composites (Huang et al., 2017) will form and serve as transitory dormitories—not jury-rigged—for those persons constructing an end-resolution to the geophysical macro-problem of Mars' final constrained para-terraformation: Richard Taylor's roofed, shell-like "Mars Worldhouse" is a particular type of "Worldship" with a star-orbiting planet, Mars in this case, at its core!

Figure 3. ***Blur Building*** operating during daytime; viewed at night, it is even more sensational. The World Meteorological Organization (established 1950), in its ***WMO Cloud Atlas***, added "special clouds" to its roster during 2017 AD, namely "Homogenitus" clouds which denote anthropogenic clouds; the term "-genitus" indicates localized factors that lead to cloud formation (Cohn, 2017). GOTO: https://www.wmocloudatlas.org. Mars' Valles Marineris has morning fog caused by crystalized water under the influence mainly of orographic effects (Mohlmann et al., 2009). All future ground-level structures built by humans could very well strongly resemble the entertaining and functional scenery pictured above. (Google Image.)

The basic premise for Geoscience's proffered theory of a past Snowball Earth state—that is, global glaciation—was enunciated first in Switzerland by Louis Agassiz (1807-1873) during his 1837 ***Discours de Neuchatel***. In its 2017 modified version, the geological concept owes much, since 1983, to Nuclear Winter theory proponents, where global wartime nuclear explosions are postulated to generate abnormal dust veils [smoke aerosols] over vast geographical regions, potentially even the whole Earth-bioshell, thus lowering our planet's global albedo, subsequently causing total glaciation and the potential for Homo sapiens' extinction (Ausmus, 1998; Dorries, 2011)! Nuclear Winter was an aberration scientifically, becoming gradually highly politicized and over-extended in both geographical scope and forecast policy impact, its initial proposers became somewhat enthralled by the world-public's attention and the popular news media's rapture. Such transformation sometimes happens with Science and, as well, Macro-Engineering (Giezen, 2012). Whilst the macro-management of gargantuan mega-projects is an ancient endeavor, 21st Century macro-projects provide challenges in modularized technological complexity and rate of assembly and/or disassembly that dwarfs any of those challenges endured by our forebears. Even as early as 1880 AD, the British novelist Percy Greg (1836-1889) equipped human travelers to Mars with waste-recycling green plants in his ***Across the Zodiac: The Story of a Wrecked Record*** (Wheeler, 2017). Some

Sci-Fi stories, for example Robert Heinlein's ***Farmer in the Sky*** (1950) and Kim Stanley Robinson's ***Red Mars***, ***Green Mars*** and ***Blue Mars*** (1992-1996) trilogy, suggest that the macro-problem of soil creation on Mars is so formidable that, in their separate intriguing fictions, soil was imported from Earth's bioshell (Pak, 2016); such so-called soil importation was actually of the exported soil's microbial content, not merely the granular mineral and freshwater matrix for those tiny invisible life-forms (Bardgett, 2016)!

About a decade ago, fake volcano building shapes were a flourishing iconic architectural form trending within Green architecture's inner-circle. That bizarre design faux-cone fad faded quickly, but macro-imagineers have persisted in their R&D efforts to tame Earth's dormant and most active volcanos, which pose harmful injury and death threats to nearby people, their livelihoods and real-property. On 12 March 2013, ***eVolo Magazine*** announced an Honorable Mention award in its "2013 Skyscraper Competition" to five architects from China for their "VolcanElectric Mask", a multi-layered tenting for Mexico's majestic Popocatepetl Volcano near Mexico City that would harvest geothermal energy! Is "VolcanElectric Mask" a huge, early inorganic format of artist-philosopher Oswald Wiener's early-1960s "Bio-Adapter" concept? Team Geographos considers Homo sapiens as Earth's nascent planet Bio-Adapters. [More on this topic in Chapter Three.] Three U.S.A. Patent Office awards are exemplary: (I) 4,738,564 issued to Thomas V. Bottillo on 19 April 1998 for his "Nuclear and Toxic Waste Recycling Process"; (II) 6,846,967 B2 issued to Larry A. Altersitz on 25 January 2005 for his "Nuclear Waste Disposal System"; (III) 7,284,931 B2 issued to Brian Stapleton Stratford for his "Magma Evacuation Systems for the Prevention of Explosions from Supervolcanoes". Long assumed by geoscientists to be geologically almost inert, Mars is now known to have endured extrusive volcanic episodes [surface lava flows] as recently as 3-4 million years ago of that celestial body's Geologic Time (Susko et al., 2017); therefore, Team Geographos anticipates that it may be possible to reactivate widespread volcanism on Mars during the 21st Century. By using technical devices such as thermonuclear explosives, comet C/2013 A1 **Siding Spring** diversion or planned ice-asteroid bombardment, Mars's crust could belch a new, thicker atmosphere that is anthropogenic in origin and, as well, serves Homo sapiens' schematic purposes related to Red Planet terraformation. This new, variant terraforming technique would simply complement the usually-prescribed Polar Zone ice vaporization scheme of those obviously layered-cap carbon dioxide dry-ice and freshwater ice sublimation Macro-Engineering/Geo-Engineering projects committed by using incidental focused solar energy or thermonuclear explosives. In effect, Team Geographos subscribes to a two-stage Mars make-over: (I) brutal partial terraforming via bombardment and nuclear explosives arriving from outer space followed by the imposition of (II) para-terraforming, using small-volume vertical dome structures and, later, Richard Taylor's large-volume pressurized atmosphere-constraining horizontally roofed structures. [Idealized true terraforming is usually defined as total transformation of a planet that is unconstrained by any anthropogenic roof covering.] Brutal terraforming techniques must be judiciously used so that ground stability problems do not subsequently plague future above-ground and under-ground structure Martian builders! Until Genetic Engineering's practitioners can medically modify human DNA to successfully adapt to Mars's early-colonization period, almost absent thin atmosphere, it might be advisable to seriously consider sending to Mars first only those persons today called Asia's "Tibetans" or South America's "Andeans" (Yang et al., 2017)!

The amateur astronomer Philipp Johann Heinrich Fauth (1868-1941) "... was one of the last of the great lunar cartographers to rely principally on visual observations..." done by telescope (Dobbins, 2014, page 703). After 1890, he issued a series of monographs laden with topographic charts of our Moon's visible features. The doublet lunar crater Fauth is named in his honor. However, during 1906 Fauth advanced the strangest geophysical notion: that our Moon was covered by a thick rind of ice enclosing an ocean of liquid water atop a rocky celestial body core. In other words, his conception rather mimics current views, post-2013, of a moon of Jupiter, Europa, discovered in 1610 by Galileo Galilei. Soon, Fauth closely associated himself with Hanns Horbiger (1860-1931) in a team-working relationship, their friendship bonded by similar views of our Solar System's icy formational event-processes. Both men believed that space was littered with chunks of ice and that stellar systems are generated by massive ice chunk collision with stars. Between 1894 and 1912 Horbiger completed his theory after much correspondence with Fauth over its details (Schroder, 1993). Horbiger alleged that about 10,000 BC Earth's gravity captured a rogue planet he named "Luna" and which corresponds to our modern-day Moon, and the capture process-event was asserted to have caused the destruction of legendary Atlantis! ***Horbigers Glazial-Kosmogoni*** (1913) was mostly penned by Fauth, but Hanns Horbiger wrote some sections of the intensively professional astronomer-ridiculed textbook. [In our experience, Team Geographos finds that after theologians and lawyers, astronomers are the most bookish intellectual slaves bound to acceptable or conventional thinking.] The reasonably reported publication fact of this, perhaps, trivial 20th Century geoscientific historical occurrence that truly intrigues Team Geographos is the evidently valid concept of the full role of space-ice in our Solar System's development. It is known, for instance, that 90 to 95% of Saturn's rings are ice by mass; Saturn's moon Tethys is 90% ice. The Horbiger-Fauth sub-hypothesis that chunks of ice from space are a source for Earthly hail as well as ice-meteoroids is substantiated by actual subsequent real-world observations (Martinez-Frias, 2000; Saul, 2006). And, even more pertinent, some researchers—all of whom have ignored Hanns Horbiger and P.J.H. Fauth's earlier 790 page publication—have proposed "Glaciopanspermia" as equivalent in validity to Lithopanspermia (Wilde and Quinby-Hunt, 1997; Houtkooper, 2011; Kraus et al., 2011). So-called "Iceteroids" surely would alter Mars' crust if they hit the ground and, even more plausible, is the prospect of modifying the Red Planet's atmosphere rapidly—enhanced warming/cooling, additional nutrients, possibly even intense precipitation process-events! Iceteroids targeted to places where volcanic eruptions may be possible are another means of amplifying anthropogenic actions planned by a program of Macro-Engineering mega-projects! It is useful, therefore, to recall the fleeting atmospheric influence that a comet fly-by had on Mars, when C/2013 A1, **Siding Spring**, closely approached Mars, **Siding Spring** even being visibly present in a camera's image of the Mars-orbiting Mars Reconnaissance Orbiter on 7 October 2014 (Farnocchia et al., 2016).

Charles Edward Housden: The First Mars Plumber

Prior to World War I, circa 1914, a highly-trained British hydraulics expert, Charles Edward Housden, calculated a complex exo-hydrology freshwater canal system that might enable, he speculated, the popularly visualized [non-Homo sapiens] Martian's tapping and storage of the planet's summertime polar ice-cap meltwater. Between 1890 and 1910 a shifting word meaning had occurred in English: a linguistic extension of the term collectively describing Mars' inhabitants: from "martial" creatures to peaceful "Martians" battling the desertification of their homeworld, a world becoming extinct by globalized drought. By empowering Martians with landscape-altering technology, some persons endowed with panoramic clarity converted them mentally into this Solar System's only world-remodeling technicians—a still unrequited skill set Homo sapiens desires! This network of canals, supposedly operated by an enfeebled (perhaps even starving) animal species, "obviously" had been constructed sometime during Mars' past Geologic Time to derive freshwater supplies from the seasonal sublimations in each hemisphere of the planet. The concept of biologically weakened, but technically accomplished [even superior to Homo sapiens] Martians had been elaborately postulated earlier by Percival Lowell (1855-1916). During the second decade of the 21st Century, Team Geographos prefers to symbolize Lowell's viewpoint by dynamic side-by-side facial imagery of an American spationaut exploring the Moon next to a post-1587 AD map, ***Fool's Cap Map of the World***, rather than bland wordage, **Figure 4**, whilst the British hydraulic specialist Charles Edward Housden's historical role can be best symbolized with the AD 1951 viewable picture in the style of Surrealism, ***Raphaelesque Head Exploding***, painted by Salvador Dali 1904-1989), **Figure 5**.

Figure 4. Fate of all human spatio-foolnauts? Are the daring, perhaps silly, humans in this picture, left-side, Edwin Eugene Aldrin (born 1930), his dichroic glass helmet face-plate obscuring his countenance, soon to become replaced by court-jester AI robots? Tele-communication from Earth to Mars is barely feasible because round-trip latencies range from 8 to 48 minutes. Common ideas of infrastructure obsolescence and civilization sustainability are means to justify change based on belief systems that preoccupy people during a definite period of time. (Google Images, modified by JMH.).

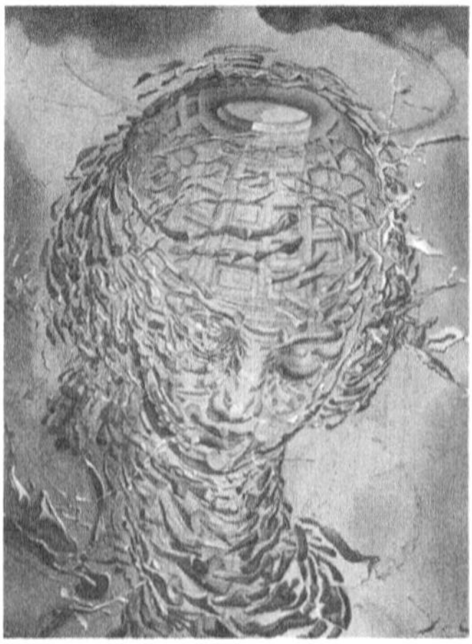

Figure 5. New symbol for the Cyborg Human Head? Homo sapiens are still collectively trying to assess the state of our Universe and Mars specifically. Dali intended his painting of a fragmented Classic human head to illustrate that human reason had been shattered by World War II's nuclear fission bomb explosions. (Google Image.)

Figure 2, above, is an oblique orbiting satellite's transmitted natural color image of Mars's North Polar Cap that gives viewers an almost God-like view of a vast quantity of stable freshwater ice awaiting near-term 21st Century future technology-targeted resource exploitation by a thriving, socially and technically progressive transplanted Homo sapiens! The shape of the most visible polar ice-cap in **Figure 2** indicates that it is primarily composed of freshwater ice and some layers of frozen carbon dioxide, all with a volume of 1.2 million cubic kilometers. In comparison, the volume of the Red Planet's northern ice-cap is less than 50% of Greenland's and only about 4% of Antarctica's! [The red-orange color of Mars is caused by the presence of oxidized or "ferric" iron in the form of red hematite (Fe_2O_3).] Four billion years ago Mars had water flowing and pooled on its surface and underground—twenty million cubic kilometers of it. Ignoring the unmapped groundwater deposits, today only about 13% is accounted for by Polar Zone ice-caps, so possibly 87% has been lost to "outer space"—that particular noun, by the way, antedates 1874 AD in English-language usage. Just like Earth today, Mars currently loses various gases from its atmosphere from the combined powerful impact of the starry Universe's impinging cosmic radiations and our immediate Sun's impactful energy emissions—space weather must be reckoned with someday, somehow.

In his rudimentary technical brief, ***The Riddle of Mars*** (1914), which is delightfully devoid of verbally hideous techno-babble, Charles Edward Housden inadvertently, and quite imaginatively, contrived a macro-planner's outline for a future Mars humanization macro-project of complex planning, construction and sociological elaboration. [Freely Internet-available at http://archive.org/details/riddleofmarsplan00housrich, his 69-page book is an excellent educational-download.] Using his honed, practical real-life hydrological skills, Housden devised a Macro-Engineering-scale scheme essential to successfully harvest the presumed hemisphere-altering seasonal periglacial runoffs. He also estimated, by calculation, the necessary energy budget that would be required to transport liquid freshwater from Mar's North and South Polar Zones, when seasonally appropriate, in order to facilitate precision irrigation pipeline agriculture, and to keep city reservoirs topped-off, with farms and cities sited in the planet's Equatorial Zone. His statements and mathematics pertaining to the "riddle"—the difference in the precipitation [snowfall and rainfall] of atmospheric water vapor in Earth and Mars—do have the tone of modernity that 21st Century Space Age expertise displays.

The only major insufficiency of Charles Edward Housden's macro-plan, from our later Macro-Engineering perspective, was his very cursory early-20th Century examination of the general nature of any global Technological Fix. All technological fixes must embrace or embody the cause-effect connecting the macro-problem to the mega-project; the proposed macro-project's benefits—that is, the "fix" being suggested—must be obvious to all those persons that will be affected by its completion, and the Geosciences are Macro-Engineering's best underpinning when the planetary database is reliably validated prior to the macro-project's inception and description. Humanity is aware of these rules because of the current worldwide public debate on the allowable, and ultimately permissible, range of Geo-Engineering schemes being devised to combat Earth's global climate regime change. Nowadays, Homo sapiens can induce disasters almost anywhere! 21st Century people share global infrastructures—in a few instances, even interplanetary infrastructures—and, thus, more and more the need to devise efficient and effective institutional, organizational and technological capacity to operate as a collective biological unit [Homo sapiens] in real-time, or selected time, in an

Earth-bioshell scale arises (Edwards, 2006). The challenging Macro-Engineering profession is ever-changing, with emerging and disruptive technologies [R&D] constantly raising new Geographical and Areographical possibilities by overtaking yesterday's exploited technologies. 21st Century people accept that machines must run our societies efficiently. Whilst design by a committee may not be the most pleasant organization of work because no single person will ever take ownership and hence pride in the accomplishment, there is a need for brainstorming when a particular macro-problem must be vanquished! Team Geographos loathes mind-numbing business buzzwords (Scudellari, 2017), instead we adore the "Art of Tinkering" style approach to job completion that is inspired by unadulterated human curiosity: 21st Century powdered three-dimensional material printing machines are the perfect means to help modern-day tinkerers express their Macro-Imagineering conceptions in physical model format! Earth's bioshell is becoming a Facility Unitized Nature (FUN), a kind of unstoppable Spaceship Earth juggernaut—from the AD 1600's, a huge wheeled wagon, under which devotees were crushed in sacrifice, the vehicle bore Krishna's image drawn annually in procession at Puri, India—not too dissimilar to "The Sultan's Elephant" mentioned in Chapter One paraded through London, England's streets during May 2006. Mars may become a FUN operated regularly by the next human generation—that is, by 2050-2100 AD. Mars will be to the coming generation what Saudi Arabia's trackless arid "Empty Quarter" landscape and the saltwater vastness of the tiny island-studded Pacific Ocean was to past human generations (Ward, 1989). A harsh environment, even in Sci-Fi stories, is a common literary means of imaging our species' future based on the ways we recall the past.

If someone who lived on Mars, experiencing a constant gravity field far less strong or noticeable than Earth's, just 38%, were to impulsively visit the Earth-bioshell, the transition would be quite unnerving: the increased constant gravity-field would affect spatial orientation, head-eye and hand-eye coordination, balance and posture, locomotion and cause some motion sickness! ***The Space Between Us*** (2017) film was set on Mars and Earth of AD 2033—Hollywood's Mars-born teenage boy meets Internet pal teenage Earth girl—but other Earthling civilians may be able to visit a simulated **Mars World** fun-palace in Las Vegas, State of Nevada, U.S.A. circa AD 2021, if Mars World Enterprises, Incorporated garners enough investors. [GOTO: http://marsworld.com.] Mars' full terraformation cannot be done Hollywood film style in a mere five-minutes [***Total Recall*** (1990) and its same title 2012 remake]!

On Earth, human-made canals were first excavated for purposes of irrigation and drainage, and their use for navigation followed. It is quite natural that Housden should attempt this daunting task, since he lived during an unnamed historical decade when economically valuable shipping canals (Suez, Panama, the Dortmund-Ems, Kiel, Manchester and Corinth channels) were dug, thereby changing the movement of large quantities of freshwater and seawater as well as the traffic patterns (routes) of capacious ships laden with bulky freight, valuable fluids and passengers. Because the Suez and Panama canals markedly reduced voyaging times between distant major seaports, the "Age of Sail" terminated. Ships moved by propellers spun by steam-engines had become available and such vessels could steer and maintain a true "sailing" course. One of the named Martian canals, "Agathadaemon", apparently corresponds to Valles Marineris, a deep highland valley located near Mars' Equator. **Figure 6.**

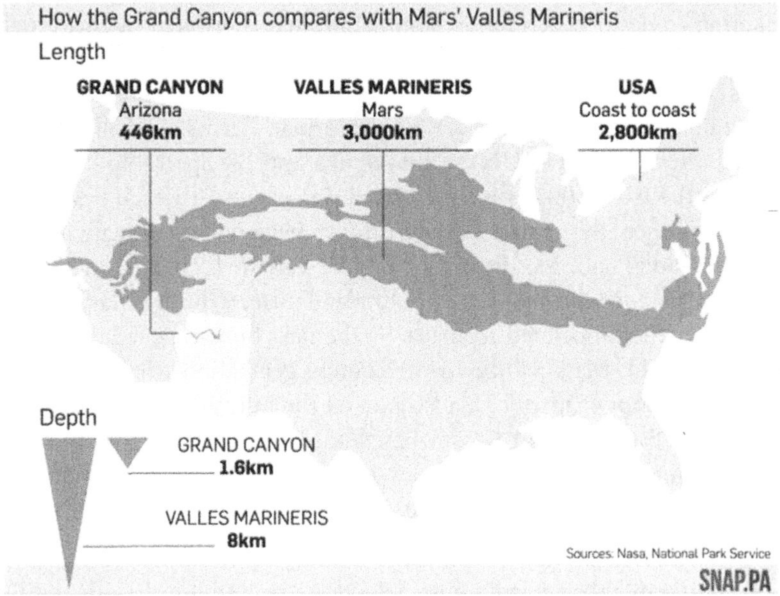

Figure 6. Valles Marineris, a truly enormous trough spanning a part of Mars's equatorial region. (Google Image.)

Noteworthy is C.E. Housden's Macro-Engineering plan of a continuous line of connected pipe segments (installed on the ground's surface or shallowly buried), supplying potable water to consumers, the purpose of which is to convey freshwater from one region to another without causing landscape erosion and reducing the volume of water wastage by evaporation. **Figure 7**. In a sense, pipelines imitate the surface and groundwater movements in catchments and could cause anthropogenic climate change if they are unsealed. Humanity's very first really long-distance freshwater pipelines were being built in aridic, but fast-developing Australia, and the U.S.A.'s "Wild West". For example, the 530 kilometer-long Goldfields Water Supply Scheme of Western Australia was started by 1896 and completed during 1903, and the First Los Angeles Aqueduct was laid over a distance of 358 kilometers from 1905 until 1913. Rather oddly, Team Geographos asserts, C.E. Housden failed to perceive that

Mar's periglacial geomorphology (SEE: Huggett, 2017, page 316-334) would be difficult landscape even for practiced Martian canal builders, and pipeline layers as well, to overcome successfully since that type of landscape imposes a highly variable pipeline/open canal permafrost interaction. When permafrost is thawed, and thus stripped of its load-bearing capacity, any poorly ballasted steel or concrete length of pipe becomes buoyant and, thus, displaced in ways that can damage the necessarily contiguous infrastructure (jacking, material stress accumulation in rigid pipe segments). That is the main reason Team Geographos will discuss, in another section of this chapter, the use of wide-diameter tensioned textile hoses laid sub-aerially conveying cold water long distances is our best recommendation to 21st Century Mars colonizers.

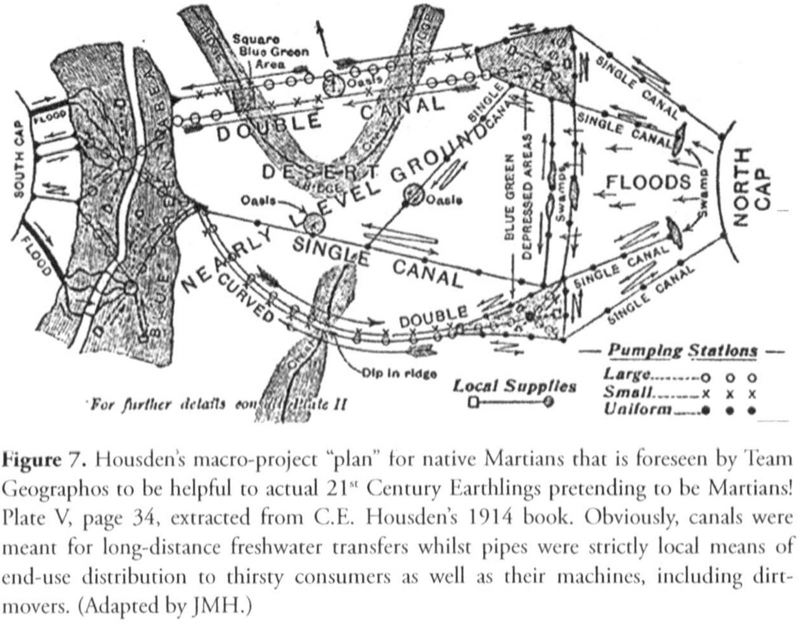

Figure 7. Housden's macro-project "plan" for native Martians that is foreseen by Team Geographos to be helpful to actual 21st Century Earthlings pretending to be Martians! Plate V, page 34, extracted from C.E. Housden's 1914 book. Obviously, canals were meant for long-distance freshwater transfers whilst pipes were strictly local means of end-use distribution to thirsty consumers as well as their machines, including dirt-movers. (Adapted by JMH.)

Earth's interconnected railway systems, the "Age of Railways", succeeded the advent of cheaply-manufactured steel and concrete; railways, in their present form, made their first appearance at the beginning of the 19th Century in British mines. During 1869, the year the Suez Canal opened to international shipping, the U.S.A.'s first transcontinental railway was completed. Railways have been proposed for our nearly atmosphere-less Moon (Shrunk et al., 1998) but adapted Earth desert and beach-motoring perfected highly maneuverable dune buggies, equipped with three-dimensionally printed, AI-enhanced, self-healing spherical wheels, with deep-tread ground-contact patches, powered by solar-power motors might better meet the everyday needs of both pressure-suited explorers (surveyors) as well as hose-laying human and networked spherical AI robot work-crews [landscapers directed by architects?]. Perhaps even future freight-carrying capsules such as Hyperloop pods, propelled by pressurized flowing freshwater, could be sent through wide-diameter hoses for timetabled resource deliveries! The exotic Dune Buggy cabbed vehicles Team Geographos envisions for everyday use for Mars surface travel will be pressurized with comfortable air-pressure, humidity and warmed cabin interiors. [The first pressurized airplane cabin became available in 1940 with the first factory prototype [test model] rollout of the Boeing 307 **Stratoliner**; interestingly, a typical two micron size bacterium's cell membrane is four nanometers thick for an aspect ratio of 1/500—about the same aspect ratio as the thickness of the exterior wall of a pressurized jet-plane's fuselage related to its interior diameter.] Mars Dune Buggies probably should contain living, sleeping and working facilities that are ultra-reliable in terms of occupant safety, and possibly much more luxurious than the sealed pressurized train carriages of China's mostly tourist-patronized Qinghai-Tibet Railway (Zhang et al., 2010; Qin and Li, 2011). At first, such Dune Buggy-style living and working facilities would perhaps be likely to resemble slightly more elaborate, but still rude, Green-designed mountain construction site huts (Goymann et al., 2008). Departure/arrival schedules and accurate cargo manifests are a requirement for optimal operation and each vehicle will be appropriately designed and constructed to meet the survival and comfort needs of all hermetically encapsulated travelers, whether they are Tibetans, Andeans or any other group of humans!

Team Geographos' Mars economic development macro-project plan, therefore, leaves plenty of open, wild, and possibly sacred Mars landscape to become deliberately dedicated planetary parklands (Cockell and Horneck, 2006). However, we exclude Valles Marineris from serious consideration as a future park; as Earthly macro-imagineers we have long indulged in Blue Sky Thinking, but the multiple mega-project terraformation effort to create Mars Redivivus—Black Sky Thinking?—we are struggling in entirely unfamiliar planet-scale territory! TAMP (para-Terraformed Arid Mars Plumbing) Macro-project will temper some of Mars's regional environments, inducing a FUN (Facility Unitized Nature), not a malicious tampering. The Red Planet ought never to be apprehended by Homo sapiens as "unspoiled" or "pristine" because, fundamentally, that assessment is an aesthetic view masquerading as a prohibitive nutty Green religious injunction! Mar does not have "canals" as Housden speculated—in fact, it does not even have eroded Nazca Lines glibly "explained" by the notorious Mr. Erik von Daniken who attributed western South America's Nazca Lines to aloof Alien visitors! Nevertheless, someday Mars may be decorated by anthropogenic artworks (Philippe, 1996). Already what is today considered junk has accumulated on Mars's ground. Because of technical problems, some human space probes burned up in Mars's atmosphere during entry which is a form of Destructive

Art—sculptural pieces that become, by physical accident and human error, self-eliminating Nature-consumed artworks. Indeed, if terraformers use thermonuclear explosives and intentional asteroid/"iceteroid" bombardment to prepare Mars for near-term future migrating human dwellers, that may also be termed Art! Team Geographos imagines Olympus Mons, the tallest volcanic mountain peak on Mars, might become seeded by Nanotechnology's Smart Dust, tiny sensors as part of an avalanche warning system perhaps (Dance, 2016); Team Geographos envisions the possibility of coordinated settled Smart Dust swarms performing as singular computational "fabrics" situated atop some Mars landscapes that require constant monitoring—effectually offering colonizers a sentient artificial coat for selected landscapes, such as the steepest slopes of Valles Marineris.

A Brief Biography of C.E. Housden

Charles Edward Housden was born 9 July 1855. He was formally educated as a civil engineer at the Indian Civil Engineering Colleges. By October of 1879, he was appointed to the Public Works Department of the Government of India, then still a part of the British Empire, having passed a vernacular examination at the departmental standard (as well as a colloquial examination in Burmese). Circa June 1903, he was serving as a sanitary engineer to the Government of the City of Rangoon [renamed Yangon in 1989 by the renamed ecosystem-nation of Myanmar]; 1908 saw him engaged as Superintending Engineer and Sanitary Engineer of East Bengal and Assam, stationed at Shillong. Whilst resident there, he helped plan and execute recovery of the region's infrastructure by repairs made necessary by the tremendously destructive Great Assam earthquake of 12 June 1897, which uplifted the Shillong Plateau by more than ten meters during a few tumultuous moments and caused widespread liquefaction and flooding of the Brahmaputra River floodplain downstream (England and Bilham, 2015). Three types of seismic waves caused by that epic event, which registered magnitude 8.1 on the Richter Scale piqued the curiosity of Richard Dixon Oldham (1858-1936), who later recognized first the distinctive presence of the Earth's core in the seismic records of the Great Assam tremblor! Although Janus was the ancient God of doorways, despite his personal experience at Shillong, C.E. Housden oddly did not, in his work about the vast speculated ancient Martian civilization infrastructure, even offer such deadly shaking threats to such infrastructure from instability of Mars' crust ["Marsquake"] and, further, he certainly had no expressed idea, framed and described verbally or by illustration, of sudden natural space debris impact landscape disruptions, the ejecta fallout pattern from each is complicated by the Coriolis Effect (Loaiciga, 2007) and ambient wind. "Marsquake" is an operative word probably not coined until circa AD 1977, after the **Vikiing-2** Mars Lander had settled slowly onto the red-orange colored ground 3 September 1976 at Utopia Planitia, and subsequently reported the first detected temblor to distant Earthlings. Future exploitation of Mars' northernmost ice-cap, thereby unloading the adjacent periglacial landscape may instigate the anthropogenic Marsquake process-event—these expectable major landscape vibrations must be anticipated as they might otherwise seriously affect settler infrastructure functions! Too, meteorites abound on Mars since its thin existing atmosphere provides little frictional hindrance to the fall of debris from outer space. If new human inhabitants, equipped with outer space-based and/or ground-based anti-bolide technology were actively present, and had a vital interest to prevent Nature's rocky space debris bombardment event-process, then major Areographic cataclysms may not ever again occur after colonization, the existential risk ameliorated at reasonable economic cost on a long-term organizational basis. In that instance, Mars' crust holds the long-term Homo sapiens-defined and demarcated status of a protected [preserved] and conservatively exploited litho-technical system of crust-rock formations, sediments and flowing fluid influence, ultimately to be macro-managed as a catastatic structure by brave, brainy, technically well-equipped, healthy settling human colonizers aided by sometimes independent AI robots. Beautiful vistas viewed by happy neo-Martians afoot, traveling in land and aerial vehicles or even afloat on fluid will be maintained by park-rangers; although the media-reported rhetoric of some popular culture terraformers is that of massive modification and transcendence, those first actually settling Mars can be aesthetically sensitive to its natural, rather flat landscapes that are like Earth's landmass Nature (Willenbring et al., 2013), which will be undergoing a global transformation due to densification of the planet's atmosphere. All United Kingdom civil servants in India were, by law, issued mandatory retirement at age fifty-five, so C.E. Housden departed government service during July 1910. On 27 November 1912, Housden joined the British Astronomical Association, which had been established in 1890 to support the efforts of amateur astronomers, whilst then residing in London, England. His landmark published article, leading shortly to his pre-World War I book on Mars's optical puzzles, was published in 1913. Between 1907 and 1914, C.E. Housden published three major reference books on topics relevant to water supply and sanitary disposal—that is, pollution solutions—macro-projects as well as mega-project management. Housden died during 1921 whilst residing on England's balmy south coast in the genteel town of Hove, in East Sussex.

British geoscientist Robert Lionel Sherlock (1875-1948), who retired from a successful career with the United Kingdom's Geological Survey in 1937, respected C.E. Housden's hydrologic professionalism and valued his Mars aqueduct mega-project speculations (Dixon, 1949). Sherlock made the following insightful Areographical deduction, circa 1922: "Can we foresee the final results when all possible engineering works shall have been carried out? Is it possible that in the yet far distant end Earth will come to a state said to exist on Mars, and be covered by enormous canals from pole to equator? The canals of Mars, if they really exist, a question still under discussion, are far greater than any our engineers imagined. The battle of the Martians with Nature has been on a much more gigantic scale than Man's conflict, and yet we hear that the Martian is on the point of extinction, and of a Mars becoming totally lifeless. Even on Mars the mighty…[macro-scale]…engineering works seem merely to scratch the skin of the planet, and the final result of Martian activity in the solar system seems likely to be infinitesimal" (Sherlock, 1922, page 347). Dr. Sherlock's was one of the few attempts (before 1966) to gain an omniscient view of Spaceship Earth from the unique perspective of Homo sapiens' ever-progressing and increasingly powerful transformative mechanical power (Douglas and Lawson, 2000) following the magisterial synthesis of global geological concepts by Eduard Suess (1831-1914), ***Das Antlitz der Erde*** [The Face of the Earth] (1885-1090). Somewhat peculiar is today's common notional meaning of "the face of the earth"—that is, the phrase's use, with almost Italicized emphasis, to refer to the existence or disappearance of someone or something rather than the opposite occurrence! Presciently, Sherlock opined in his 1922 book-survey that "…there are indications that the…[Geosciences]… doctrine of uniformity [SEE: Romano, 2015] has been carried too far" (Sherlock, 1922, page 324). Earth's atmosphere gaseous content was beginning to noticeably indicate the Industrial Revolution, as well as its globalization—not only through the emission of carbon dioxide gas by fossil fuel combustion but also direct mining of air when it became possible, after 3 July 1909, to convert gaseous nitrogen to ammonia in the manufacture of green plant fertilizer. By the 21st Century, more aerial nitrogen gas was generated synthetically than by Nature in Earth's bioshell (Gallaway et al., 2002; Keeler et al., 2016). And, of course, other chemicals in common use by Homo sapiens, especially those who are urbanized, are sensibly present in the air (Bernhardt et al., 2017). People sheltered beneath Earth's air, at least according to popular media, are verging on counting their breaths (Prairie and Duarte, 2007) because of the build-up of carbon dioxide gas in our atmosphere, an undeniable build-up (by various natural and unnatural causes) that can be technically remedied. Dr. R.L. Sherlock's 1922 insight regarding strict Uniformitarianism's inapplicability to our homeworld is, more and more, strongly supported by the findings of recent scientific research (Phillips, 2007). It has always been the case that air was a geological [outgassing] and biological [oxygenation] construction, and since Homo sapiens became extant, air is becoming a semi-anthropogenic construction. [After air, of course, approved quality drinking-water is a biological necessity that currently binds all Earthly humans and will, in future, bind all human Martians together!]

Macro-Engineering is a regional and globally transformative planning profession with blossoming dual-track career paths: (I) Geo-Engineering of Earth and (II) Terraforming or para-Terraforming of other planets. Active macro-engineers foresee the near-term future economic development of an all-encompassing outer space infrastructure by technically progressing, ethical humanity (McMahon, 2016). And, this Earth-bioshell Homo sapiens-undertaken infrastructure accumulation is being fulfilled during a controversial clique-proclaimed ceno-Anthropocene period of skimpily recorded Historical Time, not incompletely scientifically-assessed Geologic Time, for human civilization, since the radical Green claque still has nearly overwhelming predominance in the popular media and still over-fills the employee rosters of consequently pathological regulatory governmental bureaucracies. Earthly aerial carbon dioxide capture and artificial geological storage as solidified CO_2 (Kriegler et al., 2013) done by people still, but possibly temporarily, confined to a single terrestrial-type planet may engender a globalized Homo sapiens carelessness feedback, according to some excessively concerned self-proclaimed Green scientists (Matsumoto, 2006) yet, to increase the bulk of Mars's atmosphere, as future-oriented big-picture terraformers realize, the gasification of Nature-stored carbon dioxide is needed. In other words, industrial skills and capacities first developed by humans living in Earth's changeable bioshell will become valuable as colonization techniques for the physically fit, strong-willed and physically adapting Homo sapiens New Martians! Some social critics and science sages have unflatteringly proclaimed our world's global infrastructure is equivalent to the infamous foundered passenger vessel ***RMS Titanic*** in that people could fail to avert the developing mining and crop production and climatic change macro-problems envisioned by some reputable scientists and technologists. The sunk ***RMS Titanic*** nautical analogy is a horrifying reference to a superlative luxury liner on its maiden voyage, lavishly designed specifically for members of High Society, moving too fast in a known ice-field during the North Atlantic Ocean's official iceberg season whilst insufficiently furnished with lifeboats—and the unfortunate horizon-scanning crows-nest male lookouts were unequipped with binoculars or mariner's telescopes! Fortunately, 21st Century Macro-Imagineering visionaries and macro-engineer planners have access to, and benefit from, the comprehensive Geographical and Areographical studies continually generated by observatories, space exploration agencies and the variously-dedicated think-tanks that can eventually assist in the safe establishment and well-organized long-term operation of Spaceship Mars.

Mars: What's it Like Still?

Since Earth's range vertically of topographic features is just 20 kilometers, the Red Planet's approximate 30 kilometer topographic variation is remarkable—extending from the plain of Hellas Basin, about 9 kilometers below Mars's official geodetic reference surface, where the annual mean atmospheric pressure is the triple-point of freshwater [6.11 millibar], to the dormant volcanic crown of the majestic Olympus Mons [+26 kilometers, the maximum height of aggregated rock physically possible, taking into account Mars's gravitational acceleration]. Mars's area, all of which is landscape, is 144,798,500 square kilometers, or almost the same of Earth's landmass area. Its northern hemisphere is quite flat and of low elevation, whilst the southern hemisphere is much more elevated and has a highly accidented terrain. In other words, as an inner Solar System planetary macro-object, Mars has a topographical dichotomy that markedly affects atmosphere circulation patterns. Both Polar Zones contain abundant pure freshwater ice deposits, as well as buried glaciers present in the mid-latitudes. [If the two ice-caps melted, with a volume of more than 5,000,000 cubic kilometers, an theoretical global ocean about 35 meters deep could exist over the planet's geodetic reference surface—a volume of liquid exceeding the seawater volume, approximately 4,200,000 cubic kilometers, of Earth's famed Mediterranean Sea.] Winds are the most continuously active granular surface-modifying factor. Windblown sand grains of Mars saltate higher, and have longer horizontal trajectories, than Earth's sand grains; some of the macroscopic grains moved are products Team Geographos characterized as a mini-"Green-glass Effect" evident on Mars's regional albedos (Cannon and Mustard, 2015; Cathcart, 2017). As for the Earth (Heistermann, 2017), hydrologists still lack a practical theory describing water flow for the full range of the still basically unknown soil properties of Mars and, even more significantly, the differences accounted for by Mars's gravitational acceleration of 3.73 meter per second squared (compared to Earth's 9.78 meters per second squared) (Hecht, 2002). Obviously, the present-day flow of water and its areomorphological actions, as well as the future unencumbered gait and stride of humans and the track of their vehicles, will become directly observable during the mid- to late-21st Century. TAMP—para-Terraformed Arid Mars Plumbed—physically demarcates the so-called framework of Mars's planetary boundaries and, in a very practical sense, Homo sapiens resident on Mars will be the equivalent of Earth's trees (Brantley et al, 2017)!

Five years before commencement of humanity's Space Age, the United Kingdom's Astronomer Royal, Harold Spencer Jones (1890-1960), maintained Mars was a naturally vegetated planet (Jones, 1952). [A convergence of social events during 1952 that eventually catalyzed the inauguration of our Space Age was stimulated by a proposal issued from the International Council of Scientific Unions to institute an International Geophysical Year to the study of Earth's bioshell; 1952 was also the year Herman Sorgel (1885-1952) died in an automobile bicyclist run-down, a fact relevant to Team Geographos' discussions in Chapter Three.] Although "planetary engineer" received a first citation circa AD 1951 in the ***Oxford English Dictionary***, on 3 January 1953, E.F. Hope-Jones (1914-????) offered humans the emerging prospect of extra-terrestrial "planetary engineering" (Hope-Jones, 1953). Imploringly, Team Geographos touts planet-scale, mid- to late-21st Century Macro-Engineering could bring an introduced and cultivated vegetation of a spatially large areographical region of industrialized Mars. It may be that 21st Century improvements in biological solar energy harvesting (Turan et al., 2016) will result, finally, in the realization of the AD 1912 "Green Dream" of Giacomo Luigi Ciamician (1857-1922), of artificial photosynthesis applied to vast arid regions of suitable planets: "...glass buildings will rise everywhere... inside of these will take place the photochemical processes that hitherto have been the guarded secret of plants, but that will have been mastered by human industry which will know how to make them bear even more abundant fruit than Nature, for Nature is not in a hurry and mankind is" (Ciamician, 1912, pages 385-394).

The most deeply indented areographically large-scale planetary landscape feature of Mars, the Valles Marineris, honors the NASA space probe **Mariner-9**'s first imagery recordation of that slash-scar-like trough. The "Agathadaemon" canal visual illusion based on Valles Marineris was caused by thoughts verbalized by some human misperceptions. **Figure** 7 a Google Image, below. Valles Marineris, whose true geological origin remains a matter of some dispute amongst scientists, apparently is a series of crust fault-bounded troughs extending east to west more than 4,000 kilometers, with a width exceeding 600 kilometers and depth of more than 10 kilometers below adjacent high-plains, not the planet's official geodetic reference surface. Active faults will inevitably cause Marsquakes; after Mars has been colonized by Homo sapiens and AI robots, these expected rocky crust-ruptures and sporadic granular material surface displacements can be contemporaneously examined by inquisitive planet-roving field-workers. Valles Marineris is more voluminous than the Grand Canyon in the State of Arizona, U.S.A. Absent still are any extant human or tele-robotic definitive and trusted field-work data backed up by careful chemical laboratory chronology assessments. So, the true stratigraphical and absolute Geologic Time ages of Valles Marineris' associated formations were unknown during AD 2017.

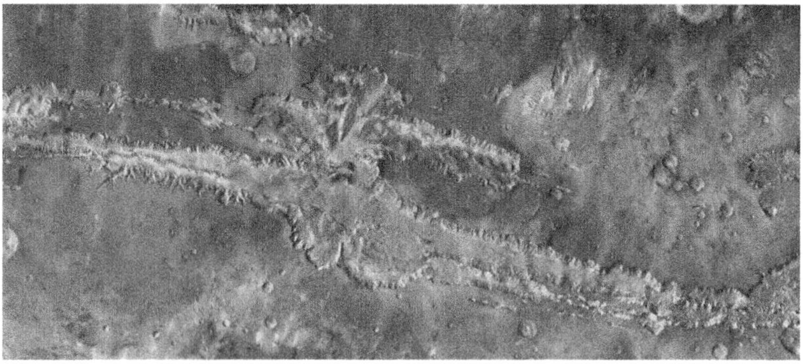

TAMPering and the Mars crust-surface, Paul Birch-style

Team Geographos' para-terraformation of arid Mars plumbing first entails a slight modification of Paul Birch's 1992 macro-project plan for the rapid plasticization and terraformation of Mars brought about by lens-concentrated incising sunlight beams focused on selected regions of the Red Planet (Birch, 1992). It seemed to us both prudent and wise to gently rework Paul Birch's soil vaporization and sintering scheme, using Mars-orbiting solar mirrors: up to 16 centimeters of surface materials could be vaporized per pass, the deadly 30 kilometer-wide beam footprint sweeping by at 270 meters per second. [Unless properly sheltered from the intense rays of the mirrored and concentrated sunlight beam, there can be no question of gawking humans looking upwards at Paul Birch's swiftly moving overhead satellites!] It becomes possible to macro-imagine Valles Marineris intentionally sealed by a deposit of Anthropic Rock, perhaps equivalent to an industrially-cast variant of Nature's ordinary basalt rock, making Valles Marineris, in some sections, watertight and, thus, precluding any significant reduction of effective rock leakage if that trough in some restricted places became useful to colonizers as artificial freshwater reservoirs and saltwater-filled waterpark places. If great quantities of water were poured into some segments of Valles Marineris, then possibly an overflow dam may have to be emplaced in Eos Chasma, at the eastern extremity of that capacious big trough. **Figure 8.** It has been confirmed that the easternmost basins of Mars's Valles Marineris contained kilometer-deep lakes "...well after the wetter, global climate optimum that characterized the previous Noachian epoch (4.1-3.7 Ga)" (Warner et al., 2013, page 675). ["Ga" is an abbreviation for "Giga-years", meaning herein 4.1-3.0 billion years ago.] These reported equatorial Mars lakes sometimes spilled over through flood channels to inundate the topography of the northern plains. [Gigantic spillovers, or outburst floods, such as that occurring at the Mediterranean Sea Basin's end-of-Messinian Salinity-Crisis, have happened in Earth's bioshell, as shall be learnt in Chapter Three.]

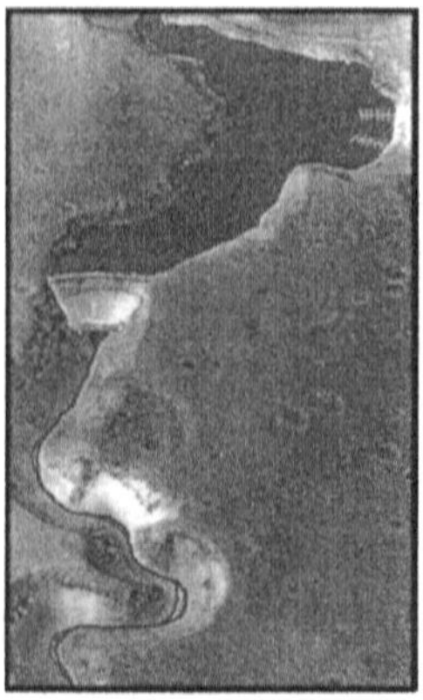

Figure 8. A NASA-parody website, now lost for-ever in the mist of past computer Atomic Clock time, illustrated a concrete arch-dam blocking the very ancient Eos Chasma natural water-overflow channel of Valles Marineris, even including marina docks for future Mars week-end pleasure boaters! Human use of concrete in Earth's bioshell

> is enormous—more than 4.0 billion tonnes annually, enough to build a wall at our planet's 12,742,000 meter-long Equator 27 meters high by 27 meters wide, a total bulk of 9,288,891,000 cubic meters of poured Anthropic Rock! Concrete, as we are familiar with, is an unlikely construction material for Mars, however. Our postulated Eos Chasma Dam can be composed of a strong, water-impermeable sail-like fabric sheet held in place by anchored super-ropes, expounded upon at length in Chapter Three for emplacement at the Strait of Gibraltar. Reservoirs do reduce dust emissions from inundated former ground dust sources—a linear aqueous sand-fence partly dividing Mars's northern and southern hemispheres! Climate change ensues (Moon, 2017).

Sampled from Earth's bioshell landmass, Sun-dried clay is little different from soft shale rock in crush strength, and the architect Nader Khalili (1937-2008) verbalized intensely on the efficiencies of in-situ material melting techniques, applied to the task of stationary extra-terrestrial human base construction utilizing hardened lava, ceramics and fused adobe. Both Khalili and Paul Birch, proponents of neo-petrology, tread in the footsteps of our ancestors in ancient Mesopotamia (Stone et al., 1998; Kopecky and Voldan, 1965)! [It is timely to reference Arthur G. Buckingham's awarded U.S.A. Patent 3,564,253, "System and Method for Irradiation of Planet Surface Area", on 16 February 1971.] During 2017, the technique of three-dimensional additive manufacturing, using a powdered Mars regolith simulant, produced bespoke printer "ink" material that could become hardened structures (Jakus et al., 2017); such "regolith-ink" can be supplemented by "bio-ink" using bacterial cultures that enables the production of numerous new materials, including aeolian sand dune bio-grouts (Lehner et al., 2017)!

In his 1992 proposed para-terraformation vision of Mars, Paul Birch suggested it could be necessary for human pioneers (and AI robots too, of course) to drill one million ground-sampling holes before phasing in his proposed de-volatilization and groove incising by orbiting satellites ever commenced. From a modern-day Macro-Imagineering and Macro-Engineering business viewpoint, Earth's largest sand sea, the Rub' al Khali, is a 770,000 square kilometer uninhabited Saudi Arabia desert of continuous aeolian sand-dunes, is an appropriate comparable landscape (Leighton et al., 2014; Stewart, 2016). Could we solidify some significant area of Saudi Arabia's Rub' al Khali (Groucutt et al., 2015), and perhaps tent only some of its tamable aridic sand sea vastness (Badescu and Cathcart, 2004; Cathcart, 1997)? Although only 0.44% of Mars's area, the Rub' al Khali is being explored for valuable deposits of oil, natural gas and various ores. So far, the costly field-work—which is constantly subjected to the pressures of time and constraining financial expenditures by our world's globalized economy—has not been a major successful economic development effort. During frequent sandstorms in that very hot and dry climate regime, human drill-rig workers must occasionally don respirators to protect their health! Worse, Paul Birch's exploration of Mars must be done a half-century before Spaceship Mars's atmosphere has been artificially generated, increased to its estimated final 330 millibar state. Mount Everest's summit is 8,848 meters; the 330 millibar air pressure that its laboring climbers encounter is at an altitude of 8,517 meters. Without supplementary oxygen, portably contained in arduously carried pressurized metal cylinders, humans must endure the condition of extreme hypoxia; hypoxia impairs a mother's ability to carry a fetus to term. Drill-rig workers living on Mars will be, physiologically, much more at risk of health impairment and actual life maintenance imperilment than Tibetans as well as those adapted natives still living full-time at high altitudes in South America's Andes Mountain Range! Like some of their visitor-brethren to high-elevation Earth landscapes, active New Martians will have to be areographers encased in a protective cocoon of electronic sensors installed in air-pressurized suits. Even assuming the rapid technical advances of lower-cost drilling, lower-cost industrial machinery, hands-on human supervision, and other efficient technologies such as drones, it seems relevant to mention that the costs of Paul Birch's Areological exploration will be very significant, and cannot but otherwise substantially affect Mars's future colonization-by-para-terraformation macro-project finish date. Mars' bioshell creation will impose many allocation macro-problems for its struggling human permanent pioneers. Furthermore, we have centered our macro-project [TAMP] on the use of lengthy tensioned-textile hoses precisely because Paul Birch's technique to emplace "air" will leave an unpredictable post-baking topography, especially since we do not wish to implement Birch's concept of Sun-scorched canal excavation and lining! Thus, there really is no practical requirement for one million explorative boreholes, any one of which could initiate unforeseeable crustal event-processes like the induced accidental eruption of Lusi Mud Volcano at Sidoarjo, East Java, in Indonesia (Karyono et al., 2017).

Generous life-sustaining requirements for a normal human worker within our homeworld Earth may total 8,000 kilograms per 365 days, of which 7,629 kilograms comprises a supply of drinking water, sanitary water, and domestic water. Adult humans can live more than thirty days without sustenance [food], but only about seven days when deprived of potable drinking water. Mars' year is 669 days, so every adult human worker stationed on Mars, laboring with just 37.5% of Earth's gravity, will need about 13,380 kilograms of freshwater per Mars year. These vital survival needs must be met prior to any other human wants being satisfied! To extract potable water from the wispy Mars atmosphere, present during AD 2017 at only 210 parts per million, could require at least 103 kilowatt-hours per kilogram of useable freshwater produced, depending on the seasonally varying relative humidity above every drill-rig worksite. Except for a few instances of exploration on the Moon and work at the orbiting International Space Station, "sustainability" practice has been applied mostly only to our Earth's bioshell macro-problems of making or earning a living exclusively in the open-air. Until mid-20th Century, air and freshwater were economic prototypes for free-goods, available in unlimited quantities with no price attached to their use. However, during the 21st Century—in what is incessantly news media-proclaimed as the ceno-Anthropocene—the Earth-bioshell has become a "scarce" commodity; scarcity, in our particular context, simply means that so many competitive uses exist for a given good that not all consumer demands for its use can be satisfied. Therefore, Mars' very rapid Paul Birch full terraformation macro-project would increase the resources available to all persons living in our Solar System. Also, by the late-21st Century, before conversion of Mars from a wild planet into Spaceship Mars is risked, it is probable that great advances will have been achieved in human tele-operation of mobile robots from safe Command Centers orbiting Mars (Landis, 2008), thus obviating the initial need for large-volume freshwater supply systems. Another program may be that which employs the Command Centers as operational controllers of the speculated spacecraft parked at L1 which might afford Mars a synthetic magneto-tail, actually an "artificial magneto-sheath", subsequently reducing the impact of space weather generated by our Sun on the Red Planet, which currently is without a natural magnetosphere, so that its thin atmosphere might cease being further eroded, possibly even becoming gradually denser because of the spacecraft's beneficial screening effect (Green et al., 2017). The Lagrange Point 1 "shadow"-inducing satellite could be made of concrete panels fabricated on Mars by material printers, then sent to the Lagrange Point 1 in outer space located between the planet Mars and the Sun. The artificial Mars "magnetosheath" may be a technology test event—that is, a small-scale prototype try-out of shield technology that might become necessary, in future, to protect our entire Solar System from Milky Way Galaxy-imposed, purely astronomical in origin, existential threats to all people (Cirkovic and Vukotic, 2016).

Mars is regularly circled by two small natural moons, the big boulders Phobos and Deimos. With a radius of 6.3 kilometers, Deimos orbits Mars in 30 hours at a distance of 23,458 kilometers whilst 11.1 kilometer radius Phobos orbits at 9,234 kilometers every 7.6 hours. These two Nature-placed rocky satellites have estimated remaining whole-body life-times of 20-70 million years until, since they are in descending spiral orbital modes, they break-up at Mars's unique Roche limit to form a ring about Mars or finally collide with that celestial body (Hesselbrock and Minton, 2017; Black and Mittal, 2015); Phobos and Deimos might be artificially hollowed to become shirt-sleeve rated TAMP Command Centers (Marden, 2002) or both could be totally capsulized by an air-tight membrane (Roy et al., 2014). Such physical enclosures imitate olden times when citizenship originally described inhabitants of walled towns! Of course, not every robot working on the ground need be directed by humans remotely: a social acclimated cadre of peripatetic AI robots could interact rationally amongst themselves to perform some essential and significant desired physical work tasks. Human civilization has already become a society of things biological and technological, sometimes linked by our world's commercial and military-owned Internet of Things!

Paul Birch's Planet Stylization Mega-project Proposal: Mars Redivivus

Team Geographos entertains the very slight possibility that life may once have lived on Mars. Therefore, we now seek to revive, or perhaps reinvigorate is a more accurate terminology, the Red Planet by introduction of humans along with their needed and wanted contingent and synthetically-enhanced life-forms: Mars Redivivus! Illumined by the Sun, Mars's solar energy receipt, under normal circumstances, is remarkably less than Earth's, only about 43%. However, currently popular books, Internet blogs et cetera describing the presumed future settlement of Mars, usually couched in the same complacent tone and idiosyncratic language phraseology as 19th Century-published guidebooks for newcomers to the New World's "Wild West", tend to promote the established communication traditions of Sci-Fi; these standardized texts laced with imaginary images often employ the descriptive metaphor of a hypothetical future civilization as a means of indicating trends and macro-problems of our own diverse two hundred ecosystem-societies! There is much honest work on a planet being settled, as well as potentially lucrative, semi-honest work in real-estate speculation and avaricious properties administration (Collins, 2008; Haqq-Misra, 2015). Today, in our Earth's bioshell, it is property developers who are building the urban "utopias". Team Geographos asks: "Why would such a common transactional situation not be the same on Mars?" And, by the way, if some visiting spationaut, aged settler or native-born Martian should die on Mars from whatever cause, the corpse would have to be cremated or undergo deliberate decomposition as otherwise it would mummify, thereby imitating some ancient burial grounds in Peru and Chile's deserts and Tibet's highest and driest elevations!

A pioneering early-21st Century textbook edited by the Candida Oancea Institute's admired and astoundingly energetic Romanian investigator Dr. Viorel Mihail Badescu, ***Mars: Prospective and Material Resources*** (2009) took a serious view of future Mars colonization. A mathematized version of the data and information presented in this book can be accessed in Chapter 26, pages 661-682, "Ecopoiesis and liquid water transportation on Mars". That still timely essay offered a complete proof for a preliminary long-distance freshwater pipeline mega-project along with a mapped first route. By circa AD 2017 humans had installed about 3.5 million kilometers of long-distance pipes—mostly to transport oil, natural gas, water—and new uses for that type of far-ranging, extensible tubular infrastructure is increasing during the 21st Century, namely to convey captured aerial carbon dioxide gas, shift biofuels from production sites to customers, restraining Hyperloop pods as well as to serve the projected our world's impending hydrogen economies!

In Chapter 26 of ***Mars: Prospective and Material Resources***, Team Geographos outlined a 4,023 kilometer-long flexible tensioned-textile pipeline extending from Mars's thick northern hemisphere Polar Zone ice-cap to a place equally distant from the touchdown sites of the NASA sent **Viking-1 Lander** (1976) and the **Mars Pathfinder Lander** (1997), ending near Valles Marineris' northern precipice. Our first-effort long hose design was realistically envisioned to harness incoming, but unfocused, solar-energy with its efficient photovoltaic upper-surface laminated coating in order to power electric water-pumps using 117.4 Megawatts. **Figure 9.**

Our single fabric pipeline's selected flow rate is 680 kilograms of freshwater per second, enough to suffice for 3,000,000 persons at a human-appreciated, basic Standard of Living. But, the reader surely recalls, ultimately the team plans to populate Mars with 2,000,000 comfortable human dwellers, leaving a significant water supply safety margin! Since the water molecule, H_2O, links 11.19% hydrogen and 88.79% oxygen, freshwater ought to be electrolyzed (delinked) to yield hydrogen and oxygen, gases desired by ambitious colonists in the midst of industrialization (Landman et al., 2017; Murukesan et al, 2016). In Earth, the highest latitude perennial spring yet known for our homeworld's Northern Hemisphere issues 520 liters per second of water and has been dubbed Ice River Spring, on Ellesmere Island, Nunavut, Canadian High Arctic; in Mars, it is probable that freshwater will have to be melted from ice technically, using embedded solar-electric heaters, before its collection and long-distance transportation within the linear fabric conduit that can be shifted (just like conflagration emergency firehoses!) to serve any situational topographical needs.

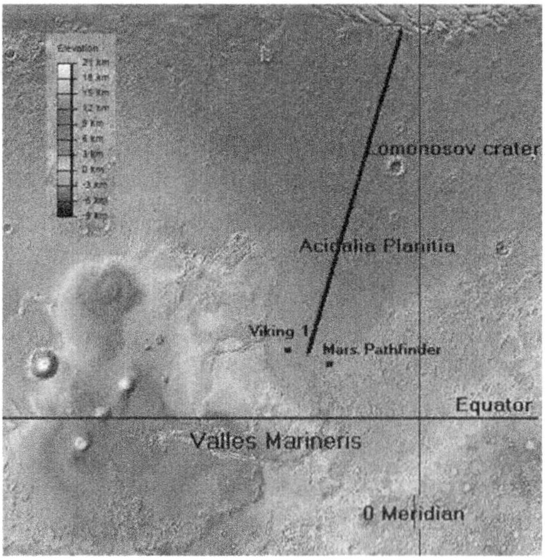

Figure 9. Solar-powered long-distance tensioned-textile freshwater delivery hose designed by team Geographos (members in Burbank, State of California, U.S.A. and Bucharest, Romania).

Whenever and wherever the prime Mars exploitation base camp is established, that place will become the equivalent of a capital city on Earth. Earth's geographic North Pole was first reached by courageous American men during 1909, but two years earlier an American zoologist, Edward Sylvester Morse (1838-1925), had pertinently asked: "How long would it take New York City to decide, in case of water famine, to tap the Great Lakes to the north, or to establish pipe lines to the [N]orth [P]ole, if it were necessary to go that distance for water?" (Morse, 1907, page 115).

Coming Attraction: Mars Anthropic Rock Landscapes

Anthropogenous lithogenesis is a historically new event-process that first occurred within our Earth's bioshell—nothing lasts forever, but manufactured rocks (concrete, ceramic, glass, brick or Paul Birch's plasticized sintered Mars ground) can endure for a long period of Mars' future existence, a possible Richard R. Landers (1919-????) "Dyboshell" comprised of mobile entirely AI robots, AI robots plus biological life-forms operating as a single harmonious working group, or individualized melds of such creatures such as cyborgs, working harmoniously on the same task. Such a Homo sapiens-inhabited "Mars Dyboshell", prior to our Earth's, may become celestially unique in our Solar System, the so-called ceno-Anthropo-Robotocene Mars Geologic Time period. Chapter Three will further elaborate on R.R. Landers' cyborg landscape concept at that relates to Mr. Herman Sorgel's much-critiqued ATLANTROPA MACRO-PROJECT of the 1920s through early-1950s, first espoused during the early-1960s when robotics research was still quite underdeveloped—that is, before widespread automated and robotized industrialization had blossomed.

Paul Birch's brutalist terraformation technique, if implemented as comprehensively proposed by him, may result in vast Mars territories that are sealed from the planet's changing ambient atmosphere, exactly like our concrete and asphalt-covered Earth bioshell's urbanized regions of roads and buildings enshrouds the landscape from our changing air and precipitation (Sutton et al., 2009). Sintering will heat Mars's regolith which then will radiate heat to the artificially thickened Mars atmosphere. Greenhouses and the Homo sapiens base camp on Mars—urban locales far more difficult for humans to enjoy daily than the early days of, say, Brazil's capital of Brasila—may be composed of glass units made from local materials (Lansdorp and von Bengston, 2006). Since ice is sometimes categorized as a form of rock, even frozen freshwater may become a competent and cheap building material applied by automated construction robots, possibly as a non-proprietary mixture of water and various strengthening fibers, into something known as Pykrete. Freshwater ice becomes as hard as brick, with a tensile strength between 344 kilopascal and 1241 kilopascal at temperatures between minus 50^0 and minus 30^0 Celsius and salinity below

0.1%. During January 2017, the NASA awarded First Place in the first round of its "3-D Printed Habitat Challenge Design Competition" to a collaborative team from Space Exploration Architecture and Clouds Architecture Office for their Mars **Ice Home**. **Figure 10** (NASA Image.).

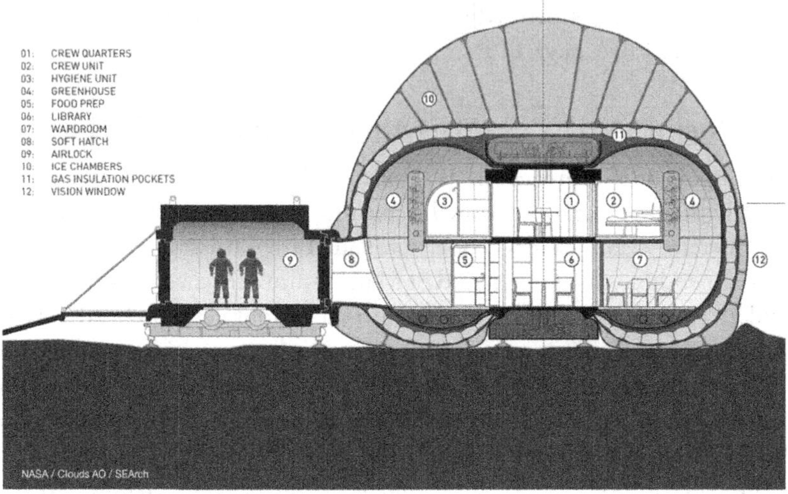

A thick walled, specially prepared ice dome offers safety and comfort to human explorers of Mar: (I) ice can shield biota from galactic cosmic rays which cause neural circuits in the human brain to lose integrity, presenting obvious symptoms of dementia (Parihar et al., 2016); (II) AI robots can assemble a home before the earliest colonists arrive and (III) ice is translucent, saving some electricity during daytime. In addition, foamy ice particles sprayed on the surrounding real-estate property abutted by the **Ice Home** can serve to entrap dust and sand grain (Askari and Kamrin, 2016) suppressing point-source emissions, to isolate a landfill of non-recyclable materials by icy covering and to protect the bio-technology construction activity of specially introduced and bred bacteria dwelling beneath the visible ground surface of the planet Mars for any beneficial new homeworld-making purpose whatsoever (Stabnikov et al., 2016; Larsson and Kaiser, 2014; Dade-Robertson et al., 2014). Making physical geological models for use indoors and outdoors is also a useful tool to formalize the master-plan an extra-terrestrial masterpiece of Macro-Engineering, the successful terraformation or para-terraformation of Mars (Hasiuk, 2014)! Earth is, culturally-speaking, a "used" planet (Ellis et al., 2013) whilst Mars is "un-used" yet by humans, only by our remote-controlled robots, with some artificial intelligence capabilities, which have left tread-marks and some discarded integuments and inactive subsidiary machines on the Red Planet's nearly vacant ground surface.

Dr. James Ross Underwood (1927-2012), a brilliant American geoscientist, was the first to propose technogenic rock as a Fourth Class of Earthly rock categorization (Underwood, 2001). **Figure 11** (RBC Image): Dr. Underwood's innovative "Anthropic Rock Cycle Diagram".

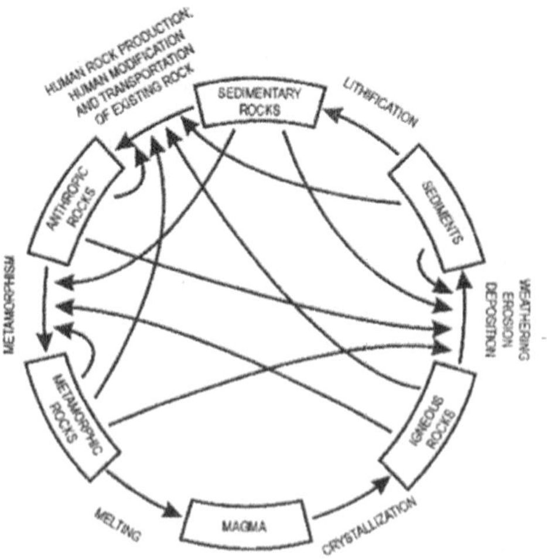

Dr. Underwood's practical Anthropic Rock Cycle proposal recognizes, at last, the pervasive geographical spread of humans and their industrial products (Cathcart, 2011). His revelatory concept came not a minute too soon since the NASA and ESA workers have publicized plans for the use of in-situ resources, such as brick as well as the stereolithographic printing of vast photovoltaic carpets, by spationauts settling the Moon and Mars! Underwood's theoretical innovation is a logical extension of the near-constant redefinition event-process inherent in Geoscience's progress. Paul Birch macro-planned construction of a network of canals on Mars using an orbiting solar lens that could melt regolith, simultaneously causing canal incision into Mars' dry landscape as well as water-impervious canal linings. Since Team Geographos suggests the use of a single tensioned-textile hose instead of interconnected Birch-style embedded lined open-canals, most of the technogenic rock ought to be made in the mined Southern Hemisphere where the highlands will be altered to create Spaceship Mars' "air". A massive display of Anthropic Rock, surfacing a vast landscape section of Mars will, quite inevitably, thrusts into the Macro-Engineering professional limelight the useful extended geographical concept that Areomorphologic phenomena are fundamentally processual. If deceased northern Africa humans can be blamed by radical Greens for promoting the presence of our present-day Earth's Sahara (Wright, 2017), then surely future humans can be awarded credit for Mars's equatorial de-desertification! Another fascinating facet of Anthropic Rock history relates to ceramic goods, specifically porcelain. Ehrenfried Walther von Tschirnhaus (1651-1708), circa 1694, worked with reverberatory furnaces that were partly heated by very large sunlight-focusing lenses, a heating technique Isaac Newton further investigated (Simms and Hinkley, 1989). During the 19th Century, the world-famous artist/painter Paul Gauguin (1848-1903) created cast material artworks that were intended to be suggestive of melting metal, symbolically emulating fluidity of matter. And, whilst Paul Birch, in 1992, wished to form [by scoring] Martian canals with laser-like beams of focused sunlight—perhaps termed by planetary stylists [professional terraformers] as a spin-off of "planet tattooing" or "planet body-art"—Team Geographos foresees no actual need for such scarification-style infrastructure attempts during the 21st Century. Tensioned-textile hoses will trump open-canals for efficient, safe freshwater delivery because they will not follow Mars landscape alterations—sediment shifts, stagnant and evaporating lakes, groundwater withdrawal destabilizing landscapes—and hoses do not leak as readily as massive sunlight-blasted canals!

Flexible and Extensible Tensioned-Textile Hoses Regularly Delivering Freshwater

To the Facility Unitized Nature (FUN) of Mars's human-occupied lakeshore homes

Whenever Mars's polar ice-caps begin to liquefy under an artificially densified atmosphere, the sufficiently fluid runoff component must be efficiently spatially directed and collected, a systemic detail best given to on-site macro-engineers to elucidate and build. This particular TAMP macro-project task to establish an initial Mars FUN will entail an effort akin to that forecast for tapping meltwater from melting Greenland glaciers for hydro-electricity combined to Libya's Great Man-Made River ecosystem-nation freshwater distribution! The total volume of freshwater meant to be delivered to a founding human settlement, the base camp, situated in the eastern part of Valles Marineris, depending on net overall crop irrigation needs and on application efficiencies [crop drip-irrigation]. It is anticipated by Team Geographos that any base camp will be surrounded by a significant terrain that had previously been sealed with Anthropic Rock for dust suppression and reduction of some jagged terrain features that would have been hazardous to fast-rolling vehicles and slower strolling humans.

Because of its elemental composition, Mars' ground can provide the necessary manufacturing for concrete, Plexiglas and plastics. Tensioned-textile hoses, fabricated on Mars, of the one-meter diameter prototype we wish to utilize must be composed of construction elements that can resist, for a considerable period of time, high internal pressure and water-freezing external temperatures. So, Team Geographos duly proposed almost a decade ago a novel textile composite material which, apart from its extremely high design flexibility in regards to shaping and the load-adapted reinforcing arrangement, specifically offers a TAMP Macro-Project administrator considerable monetary savings and, thus, forms a cost-effective Mars freshwater pipeline. A flexible conduit woven of super-strong commercially-available fibers ought to be supplemented by the introduction of super-strong carbon-nanotube fibers, or some superior material, that easily resists cold and corrosion. Our Macro-Imagineered tubular pipe for Mars should consist of overlapping layers of sealing and strengthening materials so as to resist tensile, compressive, and axial forces, all in order to properly serve as a surface-laid pipeline carrying pressurized freshwater pushed by electric pumps energized by incident solar energy. And, unlike open canals, it is improbable that unintentionally imported bacteria, deprived of sunlight, could thrive as encrustations inside one or more leak-less freshwater supply Mars hoses (Krinsley et al., 2017).

If our flexible and extensible initial freshwater hose is extended up to the northernmost precipice of Valles Marineris, then our 2,000,000 colonizers could homestead on the shoreline of slowly-filling "Lake Marineris" (indeterminate volume) fed by excess water delivery by the first hose supplemented, perhaps, by a dedicated second freshwater transporting hose! No picturesque waterfall like Niagara Falls! Unlike Richard L.S. Taylor who earlier proposed his Worldhouse solution to Mars's terraformation: namely, a quasi-global roofing of that planet with "...the greatest engineering and architectural challenge yet undertaken by humankind" (Taylor, 2001, page 236), Team Geographos prefers to canopy ideal deep-trough regions with extensive translucent suspended fabric/film. **Figure 12** (Google Image). The sky over the tenting must be almost entirely manufactured atmosphere. [Such enclosures would resemble the glass-panel covered Crystal Palace hall of displays (1851-1936) in London, England built as **The Great Exhibition of the Works of Industry of All Nations** (Schoenefeldt, 2012).]

Just five years afterwards, geographer R.B. Cathcart and the Serbian astronomer Dr. Milan M. Cirkovic had proposed preservation of our Earth by thin concrete encasement held aloft by magnetic force (Cathcart and Cirkovic, 2006). Adventurous architects might chose to decorate private properties with attractive pseudo-shipwrecks (Le Juez and Springer, 2015) and, possibly, multiple off-world [Earthlings] tourist-amusing iconic replicas of the fabled phony ***Face on Mars*** after that imaginary landscape feature was forever debunked by non-dismissible scientific evidence provided by the NASA! Eventually, "Lake Marineris" will become a Mediterranean Sea-like body of saltwater, since some of the ground forming Valles Marineris has large surface-exposed salt deposits. Humans perched beside "Lake Marineris" will likely feel a certain satisfaction much as the State of Utah's residents do, at least those who currently live on or very near the Great Salt Lake's shoreline. People situated on the floor of Valles Marineris will receive an additional benefit: (I) appreciably higher "air" pressure than 330 millibar to be generally found elsewhere because of the valley floor's remarkable 7-8 kilometer depth, below its sharply defining north and south Highland rims and (II) a reduced risk of incoming micro-meteorite damage to FUN infrastructure and unsheltered living persons, making it safer for children to play outdoors. [Visualize a red-orange foreground tinged by glittering watery blue shade in an improved remake of ***The Space Between Us***, Hollywood's dull and incredible [inconsistent and unbelievable fictional romance chase storyline] AD 2017 contribution to the teenage-love film genre.] However, the low pressure of the air at the bottom of Valles Marineris, far more than 600 millibar—the atmospheric pressure needed to have liquid water on the Red Planet's subaerial surface, will still promote a vigorous natural evaporation at the lake's surface, increasing the salinity of the water therein and generating ice-flake fogs that will, occasionally, envelope (entirely beclouding) the urbanized FUN installation, appearing much like Switzerland's still-extant ***Blur Building*** in Lake Neuchatel. Possibly a sparse desert cloud forest, not too unlike that present today in the Dhofar Mountains of Oman, may take ahold and flourish, tended by Martian gardeners. Just for fun it may be of practical use, or ecologically necessary, to imposition mega-floats, a few picturesque buoyant towns with clever and witty touristy toponyms, atop a single, or spatially separated, "Lake Marineris" incarnation (Callebrand, 2015).

Relevant to Team Geographos' TAMP Macro-project is the possibility of shifted imbalances of mass displacing any spherical planet's rotatory poles, an hypothesis first bruited during AD 1522 by Alessandro degli Alessandri (1461-1523). Geoscience has documented that the accumulation of voluminous freshwater reservoirs since circa 1950 AD is "...by far the largest anthropogenic hydrological change in terms of mass involved. This mass redistribution contributes to geodynamic changes in the Earth's rotation and gravitational field... "(Chao, 1995). Since flooding of the lowest elevations of Valles Marineris, and the impounding of a perennial mega-volume saline lake in a cold and dry desert climate regime, must extend over a lengthy period of Geologic Time it is, therefore, improbable that any catastrophic Chandler Wobble excitation whatsoever takes place afterwards (Zarkov and Gudkova, 2009). Whilst the subsurface crust structure of Mars—that is, its intra-crust zone of that planet's un-probed lithosphere—appears only slightly sketched in all the available-in-2017 AD geoscientific literature, modern analyses and syntheses indicate to Team Geographos' satisfaction that Mars' current rotation vector is stable. However, if a macro-problem arises that adversely affects our finished TAMP Macro-project, then we offer beforehand the potential to re-balance the planet's perturbed rotation with a terrain reduction [regolith mass removal via robotic excavation] or emplacement of a readily movable mass [water deposition?] at the initial man-made reservoir's antipode, much like balancing a spun vehicular wheel with strategically placed counter-weights. Of course, should any large asteroid, or sizeable atmosphere-penetrating meteorite, actually strike the indubitably artificial "Lake Marineris", a gigantic and overwhelming saltwater super-wave, or series of smaller tsunamis, would surely be the disastrous impact result! Mini-tsunamis—exemplified by towering beach pounding surge waves—might severely damage, or utterly pulverize, fixed infrastructure unwisely set too close to the visually-impressive artificial lake's sand and pebble-strewn shoreline. However, it could happen only if Mars's version of Earth's future Space Guard planet defense system were become temporarily exceeded in its protective capacity by some cosmic onslaught. Then, of course, there is always the chance of a mischievous Mars microbe arriving at the International Space Station aboard a returning rock-sampler space probe that kills spationauts living in the ISS, and threatens to invade Earth if not halted in its tracks—in other words, the unconvincing timid storyline from Hollywood's 2017 AD action-thriller flopped movie ***Life***. Greens might not latch onto this Sci-Fi film's theme since Paul Birch's scheme should successfully sterilize fully Mars's orange-red granular ground as his solar mirrors simultaneously etched a visible technical "footprint" on the, perhaps, superficially deadly Red Planet! It is worth remembering, however, that about 80% of Earth's dry landscape remains roadless (Ibisch et al., 2016), so this Birchean human impact on Mars' surface will be unique almost beyond words, but it will follow in time the alteration of the Moon's surface done for mining purposes to provide both energy and materials to the Earth-bioshell at industrial scale (Crawford, 2015). In fact, Paul Birch's scheme could become Homo sapiens' largest spatial interference within our Solar System of a single planetary surface, an un-replicated outdoor biological experiment (Barley and Meeuwig, 2017). To be sure, this would be Macro-Engineering!

The Netherlands' world-renown Modern-style architect Rem Koolhaas (born 1944) reveals in the following quotation one of the major philosophical outlook mega-problems nagging 21st Century Macro-Imagineering and

Macro-Engineering: "Beyond a certain scale, architecture acquires the properties of Bigness....Bigness is ultimate architecture....[Only] Bigness instigates the regime of complexity that mobilizes the full intelligence of [hyper-]architecture and its related fields....The absence of a theory of Bigness—what is it maximum architecture can do?—is architecture's most debilitating weakness....Big mistakes are our only connection to Bigness....[The] attraction of Bigness is its potential to reconstruct the Whole....Bigness destroys, but it is also a new beginning....Bigness...is the one architecture that can survive, even exploit, the non-global condition of the tabula rasa:...it gravitates opportunistically to locations of maximum infrastructure promise, it is, finally, its own Raison D'etre" (Koolhaas and Mau, 1995, page 494-516). Koolhaas's "hyper-architecture" is the same as humanity's hybridized and cyborg "Macro-Engineering" but, more importantly, Koolhaasian "Bigness" correctly indicates Macro-Engineering's near-term future spatial and temporal scale, the scope of its prospective cost-benefit analysis/synthesis work, construction and demolition some years past AD 2017 (Cathcart, 1997). In a few words, Mars' crust, its future human and AI robot-emplaced TAMP infrastructure, when it becomes a humanly useful mixture of natural and artificial components, a colossal composition that will be a new kind of global Chia Pet Nature will, in fact, becomes a global gizmo maintained for imaginative and hard-working people by semi-perfected post-Moon industrialization interplanetary logistics. Too, there is the very real possibility that AI robots may, somewhat like their human counterparts, evolve physically on Mars (Brawer et al., 2017).

Water flows in the veins and roots of Homo sapiens-recognized living organisms, as precious to each form of life as the gases they consume and the various forms of nourishment they process by ingestion or osmosis. Earth is a planet unique in our Solar System, rich with varied forms of water and uncountable life-forms, many forever to remain unknown to Science because of change during extended spans of Geologic Time. All known forms of life seem to require liquid water. If humans are to permanently settle Mars, then Homo sapiens must learn to gather, route, and consume fluidized freshwater derived from a hydrologic resource base of a very cold and arid planet. Pipelines are one technically feasible means to do so economically. In closing, understandably, Team Geographos must voice the warning that Paul Birch's enhanced solar irradiation of Mars's regolith will form an Anthropic Rock tombstone that inhumes any native or introduced cyborg microbial life forms (Fakhrullin et al, 2012) caught beneath it, but it will also form a firm foundation for human settlement of a new Solar System planetary abode. The potentially inhabitable abode of "Lake Marineris" will circulate according to Mars's Coriolis force deflection, perhaps even slightly because of deep-canyon winds riffling the watery surface. How much a natural geothermal flux may stir the waters of "Lake Marineris" is an unknown as yet.

No matter what today's belligerent, faulty-reasoning radical Greens wish were otherwise the fact—their so-called "North-South Divide"—Earth's 21st Century populace lives mostly in our planet's Northern Hemisphere; however, because of our planet's natural 42,000 year-long precession cycle, in about 12,000 years [circa 14,017 AD] the Northern Hemisphere will experience summer in December and winter in June—the natural change will happen gradually and people will have time to adapt geographically and technically, or to exit the Earth's bioshell! If Earth's current human population were equally spread over the total surfaces of both terrestrial-type Solar System Chia Pet planets, the density will be the same on both planets: today's estimated 7.3 billion persons divided by 342 million square kilometers equals approximately 22 persons per square kilometer. Obviously, there is plenty of room on Mars to run and play, even outdoors, if you can escape any clusters of individuals that might be perceived as "noisy and rowdy crowds"!

References Cited

Allison, S.D. (May 2006) Brown Ground: A Soil Carbon Analogue for the Green World Hypothesis. *The American Naturalist* 167: 619-627.

Askari, H. and Kamrin, K. (August 2016) Intrusion rheology in grains and other flowable materials. *Nature Materials* 15: 1274-1279.

Ausmus, W.A. (1998) Pragmatic uses of metaphor: Models and metaphor in the nuclear winter scenario. *Communication Monographs* 65: 67-82.

Badescu, V. and Cathcart, R.B. (April 2004) Architectural Ecology: A Tentative Sahara Restoration. *International Journal of Environmental Studies* 61: 145-160.

Bardgett, R. (2016) *Earth Matters: How Soil underlies civilization.* London: Oxford University Press. 288 pages.

Barley, S.C. and Meeuwig, J.J. (March 2017) The Power and the Pitfalls of Large-scale, Unreplicated Natural Experiments. *Ecosystems* 20: 331-339.

Bernhardt, E.S. et al. (March 2017) Synthetic chemicals as agents of global change. *Frontiers in Ecology and the Environment* 15: 84-90.

Birch, P. (1992) Terraforming Mars Quickly. *Journal of the British Interplanetary Society* 45: 331-340. [NOTE: his report is freely available to the public at http://www.paulbirch.net/.]

Black, B.A. and Mittal, T. (2015) The demise of Phobos and development of a Martian ring system. *Nature Geoscience* 8: 913-917.

Bolonkin, A.A. (2005) *Non-Rocket Space Launch and Flight.* Chapter 8, pages 164-172. London: Elsevier.

Brantley, S.L. et al. (3 March 2017) Reviews and syntheses: On the roles trees play in building and plumbing the critical zone. *Biogeosciences Discussions.* DOI: 10.5194/bg-2017-61,2017.

Brawer, J. et al. (24 January 2017) Epigenetic Operators and the Evolution of Physically Embodied Robots. *Frontiers in Robotics and AI* 4(1): 1-20.

Callabrand, V. (2015) Lilypads: Floating Ecopolis for Climatical Refugees. Pages 303-327 ***IN*** Wang, C.M. and Wang, B.T. (Eds.) *Large Floating Structures.* The Netherlands: Springer. 327 pages.

Cannon, K.M. and Mustard, J.F. (2015) Preserved glass-rich impactites on Mars. *Geology* 43: 635-638.

Castelvecchi, D. (2 February 2017) Astronomers explores uses for AI-generated images. *Nature* 542: 16-17.

Cathcart, R.B. (1997) Taming Mars with a Tent and a Tunnel: Creation of a Biosphere-City. *Speculations in Science and Technology* 21: 65-72.

Cathcart, R.G. and Cirkovic, M.M. (2006) Extreme Climate Control Membrane Structures: Nth Degree Macro-Engineering. Chapter 9, pages 151-174 ***IN*** Badescu, V., Cathcart, R.B. and Schuiling, R.D. (Eds.). *Macro-Engineering: A Challenge for the Future.* The Netherlands: Springer. 316 pages.

Cathcart, R.B. (2011) Anthropic Rock: A Brief History. *History of Geo- and Space Sciences* 2: 57-74.

Cathcart, R.B. et al. (2013) Shaped Metal Earth-Delivery Systems. Chapter 22, pages 507-537 ***IN*** Badescu, V. (Ed.) *Asteroids: Prospective Energy and Materials Resources.* The Netherlands: Springer. 689 pages.

Cathcart, R.B. (2017) *Sci-Fi Macro-Imagineering.* 235 pages.

Chao, B.F. (1995) Anthropogenic Impact on Global Geodynamics Due to Reservoir Water Impoundment. *Geophysical Research Letters* 22: 3529-3532.

Ciamician, G.L. (1912) The photochemistry of the future. *Science* 36: 385-394.

Cirkovic, M.M. and Vukotic, B. (2016) Long-term prospects: Mitigation of supernova and gamma-ray burst threat to intelligent beings. *Acta Astronautica* 129: 438-446.

Cockell, C.S. (2002) Mars is an awful place to live. *Interdisciplinary Science Reviews* 27: 33.

Cockell, C.S. and Horeck, G. (2006) Planetary parks—formulating a wilderness policy for planetary bodies. *Space Policy* 22: 256-261.

Cohn, S.A. (2017) A New Edition of the International Cloud Atlas. *World Meteorological Organization BULLETIN* 66: 3.

Collins, David A. (2008) Efficient allocation of property rights on the planet Mars. *Boston University Journal of Science and Technology Law* 14: 201-219.

Crawford, I.A. (2015) Lunar resources. *Progress in Physical Geography* 39: 137-167.

Dade-Roberston, M. et al., 2014. Radical Vernacular: Bacterial Architecture on Mars. *Journal of the British Interplanetary Society* 67: 346-353.

Dance, A. (19 April 2016) Terraforming a volcano, artfully. *Proceedings of the American Academy of Sciences* 113: 4234-4235.

Dixon, E.E.L. (1949) Robert Lionel Sherlock. *The Quarterly Journal of The Geological Society of London* 105: LXXXI-LXXXIII.

Dobbins, T.A. (2014) Fauth, Philipp Johann Heinrich. Pages 703-705 ***IN*** Hockey, T. et al. (Eds.) *Biographical Encyclopedia of Astronomers.* New York: Springer.

Dorries, M. (2011) The Politics of Atmospheric Sciences: "Nuclear Winter" and Global Climate Change. *OSIRIS* 26: 198-223.

Douglas, I. and Lawson, N. (2000) The human dimensions of geomorphological work in Britain. *Journal of Industrial Ecology* 4: 9-33.

Edwards, P.N. (2006) Meteorology as Infrastructural Globalism. *OSIRIS* 21: 229-250.

Ellis, E.C. et al. (2013) Used Planet: A global history. *Proceedings of the National Academy of Sciences* 110: 7978-7985.

El-Showk, S. (6 January 2017) Small Gulf nation aims for big splash on Mars. *Science* 355: 12.

England, P. and Bilham, R. (September 2015) The Shillong Plateau and the great 1897 Assam earthquake. *Tectonics* 34: 1792-1812.

Fakhrullin, R.F. et al. (June 2012) Cyborg cells: functionalization of living cells with polymers and nanomaterials. *Chemistry Society Review* 41: 4189-4206.

Farnocchia, D. et al. (1 March 2016) High precision comet trajectory estimates: The Mars flyby of C/2013 A1 (Siding Spring) *Icarus* 266: 279-287.

Fleming, J. (2007) A 1954 color painting of weather systems as viewed from a future satellite. *Bulletin of the American Meteorological Society* 88: 1525-1527.

French, G.J. (2003) The development of periglacial geomorphology: 1-up to 1965. *Permafrost and Periglacial Processes* 14: 29-60.

Freymond, P-P. et al. (1999) A Human Signature on Mars as a Symbol of a Future Terraforming.

Galloway, J.N. et al. (2002) Reactive Nitrogen: Too Much of a Good Thing? *AMBIO*: 4: 60-63.

Gerstbach, G. (2003) Mars Channel Observations 1877-90, Compared with Modern Orbiter Data. *Publications of The Astronomical Observatory of Belgrade*, No. 75, pages 209-213.

Giezen, M. (2012) Keeping it simple? A case study into the advantages and disadvantages of reducing complexity in mega project planning. *International Journal of Project Management* 30: 781-790.

Goymann, M. et al. (2008) Environmental Decision Support for the Construction of a "Green" Mountain Hut. *Environmental Science and Technology* 42: 4060-4067.

Green, J.L. et al. (March 2017) A Future Mars Environment for Science and Exploration. *Planetary Science Vision 2050 Workshop 2017 (LP) Contribution Number 1989*. Pages 1-2.

Grima, C. et al. (2009) North polar deposits of Mars: Extreme purity of the water ice. *Geophysical Research Letters* 36: L03203.

Groucutt, H.S. et al. (1 July 2015) Human occupation of the Arabian Empty Quarter during MIS 5: evidence from Mudafan Al-Buhayrah, Saudi Arabia. *Quaternary Science Reviews* 119: 116-135.

Hadhazy, A. (19 September 2015) Worlds of Pure Imagination. *New Scientist* 227: 34-38.

Hasiuk, F. (August 2014) Making things geological: 3-D printing in the Geosciences. *GSA Today* 24: 28-29.

Haqq-Misra, J. (2015) Toward a sustainable Land Use Policy for Mars. Chapter 4, pages 43-49 ***IN*** Cockell, C.S. (Ed.) *Human Governance Beyond Earth*. The Netherlands: Springer.

Harden, C.P. (2014) The human-landscape system: challenges for geomorphologists. *Physical Geography* 35: 76-89.

Hecht, M.H. (2002) Metastability of liquid water on Mars. *Icarus* 156: 373-386.

Heistermann, M. (7 March 2017) HESS Opinions: A Planetary Boundary on Freshwater Use is Misleading. *Hydrology and Earth System Science Discussion*. DOI: 10.5194/hess-2017-112.2017.

Hesselbrock, A.J. and Minton, D.A. (20 March 2017) An ongoing satellite-ring cycle of Mars and the origins of Phobos and Deimos. *Nature Geoscience* 10: 266-269.

Hope-Jones, E.F. (1953) Planetary Engineering. *Journal of the British Interplanetary Society* 12: 155-159.

Houtkooper, J.M. (August 2011) Glaciopanspermia: Seeding the terrestrial planets with life? *Planetary and Space Sciences* 59: 1107-1111.

Huggett, R.J. (2017) *Fundamentals of Geomorphology*. London: Routledge. 543 pages.

Ibisch, P.L. et al (16 December 2016) A global map of roadless areas and their conservation status. *Science* 354: 1423-1427.

Jakus, A.E. et al. (20 March 2017) Robust and Elastic Lunar and Martian Structures from 3D-Printed Regolith Inks. *Scientific Reports* 7: 44931.

Jones, H.S. (1952) *Life on Other Worlds*. London: English Universities Press. Page 206.

Karyono, K. et al. (February 2017) Lusi, a clastic-dominated geysering system in Indonesia recently explored by surface and subsurface observations. *Terra Nova* 29: 13-19.

Kaufmann, G.J. (2002) What if...the United States of America were based on watershed? *Water Policy* 4: 57.

Keeler, B.L. et al. (5 October 2016) The social costs of nitrogen. *Science Advances* 2: e1600219.

Koolhaas, R. and Mau, B. (1995) *B.S.M.L.XL* New York: Monacelli Press.

Kopecky, L. and Voldan, J. (1965) The Cast Basalt Industry. *Annals of the New York Academy of Sciences* 123: 1086-1105.

Kraus, R.G. et al. (August 2011) Impacts onto H_2O ice: Scaling laws for melting, vaporization, excavation, and final crater size. *Icarus* 214: 724-738.

Kriegler, E. et al. (2013) Is atmospheric carbon dioxide removal a game changer for climate change mitigation. *Climatic Change* 118: 45-57.

Krinsley, D.H. et al. (2017) Mn-Fe-Enhancing Budding Bacteria in Century-Old Rock Varnish, Erie Barge Canal, New York. *The Journal of Geology*: DOI: 10.1086/691147.

Landman, A. et al. (March 2017) Photoelectrochemical water splitting in separate oxygen and hydrogen cells. *Nature Materials*. DOI: 10.1038/nmat4876.

Landis, G. (2008) Teleoperation from Mars Orbit: A proposal for human exploration. *Acta Astronautica* 62: 59-65/

Lane, K.M.D. (2005) Geographers of Mars. *Isis* 96: 477-506.

Lane, K.M.D. (2006) Mapping the Mars Canal Mania: Cartographic Projection and the Creation of a Popular Icon. *Imago Mundi* 58: 198-211.

Lansdorp, B.M. and von Bengston, K. (September 2006) Design for a Mars Surface Habitat with Parts made from Locally Produced Glass. *Journal of the British Interplanetary Society* 59: 313-318.

Larsson, M. and Kaiser, A. (2014) Invincible Cities for the Materiomic Age. Pages 54-97 ***IN*** Rassia, S.T. and Pardalos, P.M. (Eds.) *Cities for Smart Environmental and Energy Futures.* The Netherlands: Springer.

Lehner, B.A.E. et al. (22 February 2017) A Straightforward Approach for 3D Bacterial Printing. *ACS Synthetic Biology*. DOI: 10.1021/acassybio.6b00395.

Leighton, C.L. et al. (15 October 2014) Interpreting and modelling late Quaternary dune accumulation in the southern Arabian Peninsula. *Quaternary Science Reviews* 102: 1-13.

Le Juis, B. and Springer, O. (Eds.) (2015) *Shipwrecks and island motifs in literature and the arts.* The Netherlands: Brill Rodopi. 351 pages.

Loaiciga, H.A. (January-February 2007) The Effect of the Earth's Rotation on Ground Water Motion. *Ground Water* 45: 98-100.

Marden, A. (2002) Use of Robotically Hollowed Out Moon of Mars as a Mars Space Station. *IAF abstracts, 34th COSPAR Scientific Assembly. The Second World Space Congress, held 10-19 October 2002 in Houston, TX, USA.* Page IAA-13-3-031AF.

Martin, I.W.R. and Pindyck, R.S. (2015) Averting Catastrophes: The Strange Economics of Scylla and Charybdis. *American Economic Review* 105: 2947-2985.

Martinez-Frias, J. (June 2000) Hailstones fall from clear Spanish skies. *Geotimes News*. Pages 6-7.

Matsumoto, K. (2006) A psychological effect of having a potentially viable sequestration strategy. *Carbon Balance and Management* 1: 4.

Maziz, A. et al. (25 January 2017) Knitting and weaving artificial muscles. *Science Advances* 3: e1600327.

McKay, C.P. (2009). Biologically Reversible Exploration. *Science* 323: 718.

McKinney, N.B. (2014) *Junkyard Mars.* San Francisco: CreateSpace Independent Publishing Platform. 360 pages.

McMahon, S. (2016) The Aesthetic Objection to Terraforming Mars. Pages 209-218 ***IN*** Part IV of *The Ethics of Space Exploration*. The Netherlands: Springer.

Meyer, W.B. (2004) Edward Bellamy and the weather utopia. *The Geographical Review* 94: 43-54.

Miller, T.K. (2017) *Mars in the Movies: A History.* USA: McFarland. 292 pages.

Miranda, C.A. (March 2013) Painting the Sun and Sculpting Fog. *ARTnews* 112: 60-62.

Mohlmann, D.T.F. et al. (December 2009) Fog phenomena on Mars. *Planetary and Space Sciences* 57: 1987-1992.

Moon, W. et al. (13 March 2017) A unified nonlinear stochastic time series analysis for climate science. *Scientific Reports* 7: 44228.

Morris, M.S. (2014) Galaxy Gadgeteers: Architects in Space. *Journal of the British Interplanetary Society* 67: 272-278.

Morse, W.S. (1907) *Mars and Its Mystery.* New York: Little, Brown. Page 115.

Murukesan, G. et al. (2016) Pressurized Maritian-Like Pure CO_2 Atmosphere Supports Strong Growth of Cyanobacteria, and Causes Significant Changes in their Metabolism. *Origins of Life and Evolution of Biospheres* 46: 119-131.

Osczevski, R. (April 2014) Martian windchill in terrestrial terms. *Bulletin of the American Meteorological Society* Pages 533-541.

Pak, C. (2016) Terraforming and the City. *Ecozon@* 7: 75.

Parihar, V.K. et al. (2016) Cosmic radiation exposure and persistent cognitive dysfunction. *Scientific Reports* 6: 34774.

Pasztor, A. (28 February 2017) Space Tourism Is Touted. *The Wall Street Journal* CCLXIX: B2.

Patterson, T. and Kelso, N.V. (Winter 2004). Hal Shelton Revisited: Designing and Producing Natural-Color Maps with Satellite Land Cover Data. *Cartographic Perspectives* 47: 28-55.

Phillips, J.D. (2007) The perfect landscape. *Geomorphology* 84: 159-169.

Philippe, J-M. (1996) Towards a work of art on planet Mars. *Planetary and Space Sciences* 44: 1463-1470.

Prairie, Y.T. and Duarte, C.M. (2007) Direct and indirect metabolic CO_2 release by humanity. *Biogeosciences* 4: 215-217.

Qin, Y. and Li, G. (2011) Permafrost warming under the earthen roadbed of the Qinghai-Tibet Railway. *Environmental Earth Science* 64: 1975-1983.

Richter, D. (2007) Humanity's Transformation of Earth's Soils: Pedology's New Frontier. *Soil Science* 172: 957-967.

Romano, M. (September 2015) Reviewing the term uniformitarianism in modern Earth sciences. *Earth-Science Reviews* 148: 65-76.

Roy, K.I. et al. (October 2014) Shell Worlds: The Question of Shell Stability. *Journal of the British Interplanetary Society* 67: 364-368.

Ruiz-Calderon, J.F. et al. (12 February 2016) Walls Talk: Microbial biogeography of homes spanning urbanization. *Science Advances* 2: e1501061.

Saul, J. (2006) Ice meteorites: is it prudent to ignore anecdotal reports. *Meteorite Quarterly* 12: 20-21.

Schoenefeldt, H. (1 August 2012) Creating the right internal climate for the Crystal Palace. *Proceedings of the ICE—Engineering History and Heritage* 165: 197-207.

Schon, S.C. (2009) Reversible Exploration Not Worth the Cost. *Science* 323: 1561.

Schroder, R. (1993) *Universal Ice: Science and Ideology in the Nazi State.* London: St. Martin's Press. 189 pages.

Scudellari, M. (26 January 2017) Buzzwords and big dreams. *Nature* 541: 450-453.

Sherlock, R.I. (1922) *Man as A Geological Agent: An Account of His Action On Inanimate Nature*. London: H.F. & G. Witherby. 372 pages.

Shrunk, D. et al. (1998) Physical Transportation on the Moon: The Lunar Railroad, page 347 ***IN*** Galloway, R.G. and Lokaj, S. (Eds.) *Space '98: Proceedings* *of The Sixth International Conference and Exposition On Engineering, Construction, and Operations In Space, Albuquerque, New Mexico, April 26-30, 1998.* American Society of Civil Engineers.

Simms, D.L. and Hinkley, P.L. (1989) Brighter than How Many Suns? Sir Isaac Newton's Burning Mirror. *Notes and Records of The Royal Society of London* 43: 31-51.

Sinha, N. et al. (February 2017) Survivability and growth kinetics of methanogenic archaea at various pHs and pressures: Implications for deep subsurface life on Mars. *Planetary and Space Science* 136: 15-24.

Soo, R.M. et al. (31 March 2017) On the origins of oxygenic photosynthesis and aerobic respiration in Cyanobacteria. *Science* 355: 1436-1440.

Stabnikov, V. et al. (2016) Sealing of sand using spraying and percolating biogrouts for the construction of model aquaculture pond in arid desert. *International Aquatic Resources* 8: 207.

Stewart, S.A. (October 2016) Structural geology of the Rub' Al-Khali Basin, Saudi Arabia. *Tectonics* 35: 2417-2438.

Stone, E.C. et al. (1998) From Shifting Silt to Solid Stone: The Manufacture of Synthetic Basalt in Ancient Mesopotamia. *Science* 280: 2091-2093.

Stychacz, T. (March 2017) The Political Economy of Potato Farming in Andy Weir's *The Martian. Science Fiction Studies,* #131, Vol. 40, Part 1.

Susko, D. et al. (2017) A record of igneous evolution in Elysium, a major Martian volcanic province. *Scientific Reports* 7: 43177.

Sutton, P.C. et al. (2009) Paving the planet: impervious surface as proxy measure of the human ecological footprint. *Progress in Physical Geography* 33: 510-527.

Tarolli, P. and Sofia, G. (2016) Human topographic signatures and derived geomorphic processes across landscapes. *Geomorphology* 255: 140-161.

Taylor, R.L.S. (2001) The Mars Atmosphere Problem—Paraterraforming—The Worldhouse Solution. *Journal of the British Interplanetary Society* 54: 236-249.

Thomas, P.C. and Gierasch, P.J. (1985) Dust devils on Mars. *Science* 230: 175-177.

Turan, B. et al. (7 September 2016) Upscaling of integrated photoelectrochemical water-splitting devices to large areas. *Nature Communications* 7, Article 12681. Pages 1-9.

Underwood, J.R. (2001) Anthropic Rocks as a Fourth Basic Class". *Environmental & Engineering Geoscience* VII: 104-110.

Vaidyanathan, G. (28 February 2017) Could a bacterium successfully shepherd a message through the apocalypse? *Proceedings of the National Academy of Sciences* 114: 2094-2095.

Vasavada, A.R. (March 2017) Our Changing view of Mars. *Physics Today* 70: 34-40.

Warner, N.H. et al. (2013) Fill and spill of giant lakes in the eastern Valles Marineris region of Mars. *Geology* 41: 675-678.

Ward, R.G. (July 1989) Earth's Empty Quarter? The Pacific Islands in a Pacific Century. *The Geographical Journal* 155: 235-246.

Wheeler, RM. (2017) Agriculture for Space: People and Places Paving the Way. *Open Agriculture* 2: 14-32

Wilde, P. and Quinby-Hunt, M.S. (August 1997) Collisions with ice/volatile objects: Geological implications—A qualitative treatment. *Palaeogeography, Palaeoclimatology, Palaeoecology* 132: 47-63.

Willenbring, J.K. et al. (March 2013) Earth is (mostly) flat: Apportionment of the flux of continental sediment over millennial time scales. *Geology* 41: 343-346.

Wolf, M.J.P. (2012) *Building Imaginary Worlds: The Theory and History of Subcreation.* London: Routledge. 408 pages.

Wolf, M.J.P. (2014) *LEGO Studies: Examining the Building Blocks of a Transmedial Phenomenon.* London: Routledge. 320 pages.

Wright, D.K. (26 January 2017) Humans as Agents in the Termination of the African Humid Period. *Frontiers in Earth Science* 5: 1-14.

Yaalon, D.H. (2007) Human-induced Ecosystem Process Always Involve Soil Change. *BioScience* 57: 918-919.

Yang, J. et al. (18 April 2017) Genetic signatures of high-altitude adaptation in Tibetans. *Proceedings of the National Academy of Sciences* 114: 4180-4194.

Zarkov, V.N. and Gudkova, T.V. (2009) The period and Q of the Chandler Wobble of Mars. *Planetary and Space Science* 57: 288-295.

Zhang, K-c et al. (2010) Damage by wind-blown sand and its control along Qinghai-Tibet Railway in China. *Aeolian Research* 1: 143-146.

Zhou, Y. et al. (2015) A global map of urban extent from nightlights. *Environmental Research Letters* 10: 054011.

CHAPTER THREE

Section I: Exterior Ring Cyborg Zone

CYBORG ATLANTROPA: Macro-Imagineering the Mediterranean Sea Region

Humans can willfully act independently of their needs but, as yet, extant AI robots cannot desire anything additional to basic operational mechanical needs (Rivers, 2008). Ultimately, mobile and immobile three-dimensional material printers, and the distributed additive manufacturing such machines could enable, will democratize survival of Homo sapiens, cyborgs and AI robots (Royakkers and van Est, 2015) by permitting the physical basis of interminable agency or purposiveness within spherical and shell-like spatial regions within this Universe. An unlimited mental model of "human", "cyborg" [English-language portmanteau for "cybernetic" and "organism"] and "AI robot" insures embodied intellectual change, including wisdom that is a multi-dimensional and adaptive attribute of all such designated persons with varying physiques that has an ultimate meaning which can never ever be finally defined. [The word cyborg was coined by Manfred E. Clynes (1925-????) and Nathan Schellenberg Kline (1916-1983) in their magazine article "Cyborgs and space" published at pages 26-27 and pages 74-76 in ***Astronautics*** during September 1960; unmistakably, they meant "cyborg" to be a name for a newly recognized extension of mainstream Science studies dealing with useful Space Age integration of biology, mechanics and electronics research. One of the most compleat U.S.A. Patent Application Publications for a compliant and workable cyborg, recommended for perusal by our readers, was awarded to Boris Kaplan on 30 January 2014 as US2014/0032466 A1.]

None of the three recognized main groupings of persons that we visualize ought never to be labeled as "users", "owners" or "bosses" because, it is Team Geographos' hope, they will always work empathically together, harmoniously as an eco-social unit for the general welfare of all persons; no team member will be disinherited or emancipated, only aware of being and of caring for others. Team Geographos, sincere persons one and all, likes the pleasantly homogenizing English-language term "Scienceers" for these barely separable groups of non-passive thinkers and diligent doers, the "machine-man and the man-machine" groups, as Richard R. Landers (1919-????) had deftly categorized by AD 1966, operating on the basis of scientifically correct data and information, because that term seems to capture the spirited adventurousness of modern-day Macro-Imagineering and Macro-Engineering! Landers patented an impressive apparatus which challenged the then current conception of life and artifice: "Self-Repairing and Self-Sustaining System", U.S.A. Patent 3,274,401 awarded 20 November 1961. Someday the intact Earth and Mars dyboshells will become analogues for human, cyborg and AI robot universalized spaceflight adventures—in fact 21st Century planet-sized spaceships that are technically more advanced than the caves once housing troglodytes, ocean-going boats and ships including submarines, underwater habitats, remote locale mining camps and subterranean command and control centers of various peacetime and wartime infrastructures. In other words, Scienceers will actually exist during an Astrologic Time-realized and materialized future spacial congruency of Richard R. Landers' transmogrified 1966 dybosphere concept and our Sun's heliosphere! The media over-popularized ceno-Anthropocene name for a transformational period of documented planet-wide natural and unnatural changes is almost usefully defunct, the impractical Earthly Geologic Time descriptor will soon become passé, old-fashion, meaningless and unemployable since, by 2150-2100 AD, all minded inquisitive creatures called Scienceers will collectively recognize the connected predominance of the Internet of Things within a singular Solar System-wide Machinocene!

Spaceship Earth and Spaceship Mars, of course, is a mechanistic metaphor which serves adequately professional Macro-Imagineering in any ongoing and future endeavor to technically correct, or improve, the geophysical operations of those planets. At considerable financial cost, humanity has undertaken to erase the anthropogenic gaps of our 20th Century homeworld atmospheric ozone layer, for instance (Previdi et al., 2014). Like the internationally-revered geochemist Vladimir Vernadesky (1944), optimistic members of Team Geographos expect cooperative Scienceers cognition and resulting announced solutions, whether derived from AI robots, humans or cyborgs, will present all such individual persons with viable timely solutions for 21st Century Solar System environmental macro-problems! ["The spatializing word "environmentalism", the encompassing-of-all physical things as well as ambient energies in the known Universe, dates from AD 1923; radical Green-garbled rendition of "environmentalism", with its current peculiar anthropo-phobic ideological skew, is nowadays formulaic, a term that gained impressive public popularity because it has, since circa 1970 AD, resonated with the aroused world-public's anti-pollution zeitgeist but, nearly half a century later, has become merely a buzzword, currently overexposed in the shockingly politicized world news media, and consequently has met an unpitied fate of coopted verbal status.] It is true that our species, Homo sapiens, are necessarily unaware of its entire Earth-bioshell history but progressing AI robot technology has a recorded history that has very few unfound "missing links"! During the early-1950s, Claude Elwood Shannon (1916-2001) build a small, boxy machine that, when switched "on" unfailingly reached out with an articulated "arm" to switch itself "off"; Walter Redfern, at page 33 in his 1989 book ***Cliches and Coinages***, wondered if what distinguishes Homo sapiens from our super-computers "...is our ability to say Yo or Nes..."—we make mistakes sometimes, and our only naturally effective off-switch is a physician-confirmed, possibly by suicide, brain-death! Sciencers, at least those who become self-identified specialists in Macro-Engineering, cannot but truly physicalize the Machinocene rapprochement between near-term future Machinocene geoscientists and biologists, some of whom may be as easily repairable as machines or pure machines imbued with the best pre-selected human qualities, physical and mental! In other words, sentient companion species, not Aliens produced by Homo sapiens!

Adding to the insightful and revolutionary biological findings about blood's natural circulation made by the celebrated anatomist William Harvey (1578-1657), where venous membranes were compared to pound-lock sluice gates and were present to obstruct blood reflux (Boyle, 2008), were the marvelous vividly illustrated explanations, first done by 1926 AD, offered by the European graphic designer Fritz Kahn (1888-1968). Kahn had experimented with well-drawn and painted chart-images, done by expert industrial-design illustrators that compared individual human physiques and functions to a stand-alone factory of pipes, chutes, reservoirs, conveyor-belts, chemical treatment chambers, and the human brain's circuitry to machine control panels studded with dials, buttons, illuminated indicators as well as open and closed levers. In other words, Kahn, who pioneered popular Science communication for the world-public, used a familiar object as a metaphor for known biological event-processes (Debschitz, 2013). On 20 September 1966, the inventor Richard R. Landers, **Figure 1**, was duly awarded U.S.A. Patent 3,274,401 for his "Self-Repairing and Self-Sustaining System". Landers sought the perfect machine, a machine or system capable of doing work continuously, detecting wear and breakdown within itself, able to repair the damage, and to keep working. From this pioneering patent, first filed with the U.S.A. Patent Office on 20 November 1961 and Figure 3 therein centered on a lozenge-shaped sub-assembly object Landers identified as a "dybloc". "In

Hebrew mythology there is the term *dybbuk* (dib-buk) which means 'unassigned soul.' [Born in 1971, the French artist Gregory Chatonsky hints at a machine's solitude.] Taking the first three letters *dyb* and adding the vowel *o* for... [assonance] produced the prefix *dybo*. From this, it was easy to arrive at dybology, the 'artificial' equivalent of biology" (Landers, page 13). [The 1938 Yiddish movie ***The Dybbuk*** famously told the story of diabolical interference in human affairs by an evil spirit was based on a 1917 play written by Shloyme Zanvi Rapaport (1863-1920).] In Richard R. Landers' landmark book ***Man's Place in the DYBOSPHERE: What will your life be like in a world of mechanized men and humanized machines?*** Landers' use of "sphere" more than fifty years ago was premature because it was not then phenomenally global; by circa 2017 AD it was more accurate to consider any such future region a global, and self-evolving, "dybo-shell" entity.

Figure 1. Circa 1965 AD, American industrial electronics engineer and product production administrative expert Richard R. Landers. Responsive technologies link computation and biota. (Google Image.)

Some Regional Landscapes as Happy Cyborgs

Alan Patterson Rudy (born 1962?) seems to be one of the leading originators of the cyborg geographical region concept of Earthly landscapes (Cantrell, 2016; Lokmann, 2017)—and by obvious categorical spacial extension to other Solar System terrestrial-type planets, useful for Mars landscapes—when he applied the idea to the State of California's infamous basinal desert landscape receptacle of the artificial Salton Sea, the grandly-named Imperial Valley: "The articulation of a cyborg perspective sees the environmental conditions of the [Imperial] Valley as the product of relations comprised of uneven and indeterminate ecological process, techno-scientific trajectories, and human practices" (Rudy, 2005, page 19). Contemporary technology's capacity to transcend old oppositions between Nature and human civilization is exemplified by the proliferation of cyborgs which are, basically, useful and convenient mixtures or entanglements of Nature's and Synthetic Biology's organisms and designed machines (Edgeworth, 2010). When Nature had sole custody and oversight of the Salton Trough it was then our homeworld's largest below sea-level Earth-crust depression (7,720 square kilometers). But, after 1905-1907, due to a Macro-Engineering irrigation mega-project's failure, an accidently imposed huge lake, the "magnificent" anthropogenic Salton Sea, then at that crucial strategic moment land speculators renewed their post-1901 AD positive boosterism to promote realtors' land sales to settlers by bestowing a humanizing futuristic region-landscape toponym, the Imperial Valley! The 21st Century region encompassed by Dr. A.P. Rudy's cyborg Salton Sea is laced with vast interconnected infrastructures, especially agricultural plumbing, that is in need of 21st Century redevelopment and replacement. Some of the private-sector groups of professional macro-imagineers working towards a macro-solution for the Salton Sea and its surrounding desert landscape's notorious hydro-geophysical maladjustments are: (I) Michael Clinton Consulting LLC in Las Vegas; (II) Tetra Tech in Pasadena, California; (III) Jennings & Johnson JJP in Palm Springs, California; (IV) Silverthorn Engineering in Sunnyvale, California as well as (V) Geographos in Burbank, California. The Salton Sea's water is now about 40% saltier than the world-ocean's seawater and its areal reduction in ground coverage because of evaporation creates a macro-problem of blowing particulate dust air pollution, seabird poisoning from toxic prey and fish die-offs from anoxia and heat stress (Heness et al., 2017). The nearby Gulf of California also shows indications of apparent ecosystem degradation (Paez-Osuna et al., 2017). Some significant observed signs for both the Gulf of California and the Salton Sea are indisputably related to long-term reduced freshwater flow from the increasingly arid continental interior Colorado River Basin to the Gulf of California. To make the best of a bad situation, Team Geographos recommended inundation of the former Salton Trough with seawater gravity flowing from the Gulf of California, led to the Salton Trough ground depression by navigable sea-level canal, a ship canal which would permit development of an international shipping mega-port, floating photovoltaic power-plants and associated desalination complexes producing and transporting, via long-distance pipeline, low-cost freshwater to the State of Nevada, State of Arizona as well as to the State of California's communities and farmlands (Cathcart, 2013).

Dreamt Macro-projects Testing Neo-"Old World" Nature Realities

The litmus test for any mega-project is the overall effect it could have on our globalized human civilization. Homo sapiens' claims to Earth's future Geologic Time invoked by Macro-Imagineering are made geophysical by artifacts and the inclusion of imaginary technologies and infrastructures to meet a geopolitically-negotiated goal that would otherwise be infeasible has sometimes been called "magical thinking". Because dreaming is the core of the human condition, Team

Geographos believes that the greatest goal-achieved planned mega-projects are the product of a single macro-engineer or at least a small, efficiently structured team of macro-engineers, the kernel of Macro-Engineering integrity. Human civilization's present-day condition is the artefactual result of the shared infrastructural dreams of previous generations and its prolongation depends to a great extent on accepting the constraints imposed from those materialized magical mega-projects. One of the most monumental dream ever realized was the product of a famous Hollywood movie-maker—Cecil Blount Demille (1881-1959)—who had constructed the mammoth 1923 silent-film set for ***The Ten Commandments*** that was abandoned and thoroughly buried by windy local Nature beneath the a massive drifting sand-dune of the Guadalupe-Nipomo Dunes Preserve in the State of California's County of Santa Barbara. [Filmmaker Mr. Peter Brosnan revealed the incompletely known story of Demille's copies of ancient Egyptian architecture in his treasure hunt-styled documentary ***The Lost City of Cecil B. Demille*** (2016). **Figure 2**, below (Google Image).] The 1920s was the heyday decade of big cinema-set constructions. However, only the sand-covered set for the motion picture theatrical epic remains because it was intentionally interred rather than scrapped. Robin Hood's castle in which Douglas Fairbanks (1883-1939) romped during 1922, the faked fabulous Babylonian city for ***Intolerance*** (1916) by David Wark Griffith (1875-1948), as well as the gigantic city of Jerusalem for Demilles' ***King of Kings***

(1927) were destroyed. A local legend, however, was confirmed. Similarly, excavations at the Dakhla Oasis in Egypt during 2014 that may account for the 524

BC "disappearance myth" about an ambitious warring Persian conqueror's army, led by Cambyses, that supposedly disappeared in the desert sands like Demille's volumetrically enormous movie set; in fact Cambyses' army may have been ambushed by the forces led by King Petubastis III so that instead of a Persian army group buried by a devastating sandstorm—looking perhaps like those stalwart French soldiers at Verdun during World War I who were interred by thrown soil caused by relentless German artillery shell bombardment explosions—there may be human remains, their weapons and supplies covered by desert sand on a battlefield. Petubastis II evidently only constructed a single temple building, dedicated to the god Thoth at Amheida, which was his powerbase, and from which he ruled Dakhla Oasis, the encampment which sallied forth to inflict a humiliating defeat on Cambyses. Monumentality represents a form of durable social practice—erection of hypertrophic artifacts (structures and buildings) that obviously exceed that mass pragmatically necessary to achieve human survival—that is most often manifested by powerful eco-social elites wielding controlled energy, information and matter (Trigger, 1990). Even spectacular destruction of a monument can enshrine such massive, grossly material objects to human memory in the form of verbal (oral and written) formats despite the absence of physical ruins. One has to wonder what people of the 22nd Century Earth could assume from evidence at had about Cecil B. Demille's sand-buried ghost city of 1923 or Michael Heizer's ***Levitated Mass*** (2012) artwork shown in Chapter One's **Figure 9**!

After popular-press exposure of the debated Continental Drift concept of Alfred Lothar Wegener (1880-1930), and before the mid-1960s academic and mining/oil industry acceptance of the reality of the ongoing Earth-crust's plate tectonics, two daring doyens of American Geography, Drs. Samuel Van Valkenburg (1890-1976) and Ellesworth Huntington (1876-1947) had already published their university-level textbook, ***Europe*** (1935) which accepted the origination of the Alps (O'shea, 2017) by tectonic process-events, a fact of academic history first revealingly emphasized by the noteworthy American geographer Dr. William H. McKinney (1923-2016). Shared by seven ecosystem-nations, the geological fault contact-zone named the Periadriatic Seam marks where the Adriatic tectonic plate fluxionally abuts Europe's tectonic plate. When alpine glaciers advanced or retreated during and after the Ice Age, people traversed high-altitude alpine passes, even farming some of them, to shorten their journeys (Hafner et al., 2017). Still under construction since 1999, and not expected to be finished until 2026 AD, the Brenner Base Tunnel penetrates the Periadriatic Seam dividing Innsbruck (Austria) and Fortezza (Italy); trains traveling normally at 250 kilometers per hour cruising speed will soon connect Verona (the Veneto) and Munich (Bavaria, Germany). Previously, a macro-imagineer had planned to pass floating barges over the Periadriatice Seam by stepped freshwater canal. Pietro Caminada (1862-1923) first offered his plan for such a sub-aerial/tunnel passage during 1907

AD and was awarded subsequently a U.S.A. Patent 955,317 on 19 April 1910 for his "Tubular Inclined Ship-Lock" mega-project concept. [Caminada's canal—the passenger and freight-pod Hyperloop of its day—is detailed in the English-language by an anonymous author in the 6 March 1908, Volume 1, #4 issue of ***The Engineering News*** at pages 95-96 and earlier, more cursorily, in the 16 February 1908 issue of ***The New York Times***.] Pietro Caminada's canal route was to have connected the major Mediterranean Sea Basin port of Genoa and, by way of Lake Constance, Basel (Switzerland). The early-20th Century architectural dreamer Paul Karl Wilhelm Scheerbart (1863-1915) had envisioned all-glass cities festooning the Alps—anthropogenic mountains of glass ensconced amongst the majesty of white snow-capped peaks! Morphosis Architects member Mr. Thom Mayne has designed a slender 381 meter-tall mirrored hotel, set for completion by AD 2019, in a verdant valley near Vals, Switzerland. When completed, the Trans-European Transport Network (Ravazzoli et al., 2017) will offer, instead of an [probably] economically infeasible Pietro Caminada-styled Verona-Basel barge canal, a high-speed railway of luxurious train carriages and freight wagons linking Genoa to Basel through the Terzo Valico Tunnel, expected to be finished by 2021 AD, and the 2007-completed Lotschberg Base Tunnel, is being excavated during the early-21st Century. Team Geographos carefully inserted "probably" because a 92 kilometer-long railway tunnel, bored within the seabed of the Gulf of Finland, connecting Tallinn (Estonia) with Helsinki (Finland) may be undertaken soon and Norway has announced the excavation, starting after AD 2018, of its 1.7 kilometer-long Stad Peninsula Ship Tunnel, drilled and blasted through a 300 meter-high hill, that will have an operational depth of 12 meters below sea-level, a width of 26.5 meters and a height above water surface of 37 meters. **Figure 3**, below (Google Image.)

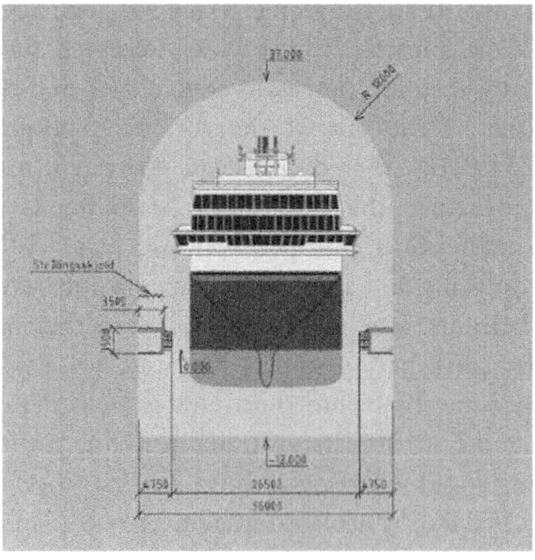

Whilst some wailing, radical, possibly even anthropophobic Green internationalists bemoan humanity's "fingerprints" [anthropogenic topographic signature

(Tarolli and Sofia, 2016)], people would not have civilization as we know it in the 21st Century—still subjected to Nature's threatening anticipated future change—if humanity had never impressed out planet's landscape with mines, roads and canals! During the Last Glacial Maximum, a massive ice-sheet coated the Alps, but today, during the 21st Century, as European glaciers recede to reveal ice-free landscapes, they may be replaced by new hydro-electric dam reservoirs (Farinotti et al, 2016; Gurung et al., 2016). In that case, humans will be doing exactly what their forebears had done: Umberto Monterin (1887-1940), then Director of the Monte Rosa Royal Observatories of Meteorology and Geophysics, during 1937, had noticed the presence of ancient disused irrigation canals at high altitude in the Aosta Valley obviously dug during a previous warm regional climate regime period when no glacial ice existed in that valley! 21st Century climate change is expected to influence the frequency and magnitude of floods, snow avalanches, debris shifts, landslides and rock-falls that can cause loss of human lives (Schwenddtner et al. 2013). Interestingly, the snow-resort of Zermatt, Switzerland, has installed a plumbed system of pipes that, desirably, allows the architected landscaping of mountain-sides with artificial snow, by an early variant of three-dimensional printing that someday could be used on Mars, so that recreation-seeking tourists will find, upon their arrival on-site at the ski-slopes, a truly dramaturgical product of Swiss Macro-Engineering outdoor stagecraft!

Reshaping an eccentric, eco-socially bumpy pre-cyborg planet Earth

By 1694, Willem Goeree 1635-1711) had drawn possibly the first printed image of the whole Earth, divided into hemispheres, without Nature's seawater content! **Figure 4**, below, a Google Image. Imagine Earth as today's Mars—that is, ocean-less!

Two noted installation artists, Adam Lowe and Gregoire Dupond, carved from a large rectangular block of polyurethane a three-dimensional topographic world map, the world-ocean's uneven basin depression represented accurately, just as the projected elevations of Earth's landmass. Publicly they announced their artwork was an artistic cartographical response to the purported [ceno-] Anthropocene's final geophysical affects, that water could be introduced to our world-ocean basin to show the overwhelming effect of a coming global rise of sea-level. The Lowe-Dupond installation, ***Antropocene Monument***, was exhibited in Les Abattoirs, Toulouse, France during October 2014. Humans and trees are alike in that, as bio-energized life-forms, they plumb the Earth-crust and project plumbed structures into our Solar System homeworld's lower atmosphere; both species of animate and inanimate life cycle huge volumes of water, fresh only in the instance of trees, but saltwater and freshwater masses in the case of human beings (Kirkby, 2016). The last sunset of mainland Europe, on the 24 of April and 18th of August, coincides with mainland northern Africa at Cap Blanc (Western Sahara), Cabo Tourinan (Spain), Land's End (England), and Tromso (Norway); Africa's "pole of inaccessibility" is located at 5.56^{0} North latitude by 26.17^{0} East longitude—about 1,814 kilometers distant from the world-ocean—whilst Spain's place of furthest access to the ocean is 362 kilometers, at 39.99^{0} North latitude by -4.51 East longitude. Why does observed regional and global sea-levels stimulate so much vociferous publicized fretting by international artists and world-famous geoscientists? Natural changes in shoreline-located places, including whatever sits atop, in or below, abutting the dynamic seawater bodies of our homeworld's ocean, such as the marble beaches of Tuscany (Nordstrom et al., 2008) or the March 2017-fall of the island of Malta's iconic Azure Window Rock Formation are inevitable and, in many instances, unstoppable. The widespread common fear and loathing regarding possible future sea-level rises, especially as these levels might be expressed locally, is generated by an incontrovertible geographical fact: settled human populations and valuable infrastructures are concentrated at coastal cities facing a tidal or non-tidal sloshing seawater mass—urban regions such as Marseille, Istanbul, Odessa (Abadie, 2016) in the case of the Mediterranean Sea-Black Sea Region. Whilst working in France, during his collage-phase, the Surrealist artist Joan Miro (1893-1983) tromped through Paris's picturesque streets loudly shouting insults to all within earshot during the 1930s on behalf of Surrealists and others, demonstrating against influential conservative political cliques that were actively thwarting their modern-day Art's advancement Miro chanted "A bas la Mediterranee" ["Down with the Mediterranean"], an uncouth bellowing meant to provoke police arrest because it indisputably indicted all prevailing Classical culture! Flash-forward mentally to 2014-2016 AD, when a traditional German science and technology museum, Munich's amazing Deutsches Museum, held its dedicated display-program "Wilkommen im Anthropozan: Unsere Verantwortung fur die Zukunft der Erde" ["Welcome to the Anthropocene: The Earth in Our Hands"]; its well-attended popular display-program was claimed by its curators to be our world's first large-scale special museum exhibition solely dedicated to the ceno-Anthropocene topic (Robin et al., 2014). There is no doubt, anywhere these days, that our Earth-bioshell is undergoing changes imposed by Nature and induced by Homo sapiens. Before AD 1929, Herman Sorgel (1885-1952) had commenced his Macro-Imagineering masterpiece, his famed ATLANTROPA PROJECT, foretold a radical induced reduction of sea-level within the Mediterranean Sea Basin caused, in part, by a voluminous emplacement of concrete in the form of dam spanning the sea strait separating Spain from northern Africa. Daniel Jutte's ***The Strait Gate: Thresholds and Power in Western History*** (2015), referenced how doors, gates and related barrier technologies that feature keys and locks continuously have shaped and reshaped the ways in which humans have understood and circumvented enclosed spaces because swinging closed doors and open doorways serve as sites of political power expression (inclusion and exclusion) as well as metaphors of salvation. Herman Sorgel, an architect and engineer by trade, wished to calm post-World War I Old World anxieties about European security in terms of food, fuel needs as well as maintenance of Europe's geopolitical independence (Hansen and Jonsson, 2014; Lehmann, 2016). **Figure 4**, below (RBC Collection) is a post-World War II image of an aged and weary Herman Sorgel. [For historical context of Herman Sorgel's ATLANTROPA, SEE: Muller, 2000; Christianson, 2012; Hansen and Jonsson, 2014; Kander and Malanima, 2015.]

This secure state of mind and body was symbolized, in many of Herman Sorgel's published, and unpublished pennings, as a placid African pachyderm situated within a bubble! **Figure 5**, below, a Google Image: the bubbled, or ensphered, Fan Yang-style "Elephant in the Room", so to speak.

In other words, a geopolitical goal generally contradictory to Joan Miro's piquant shouted sloganeering. All of Herman Sorgel's collected willed papers are thoroughly organized, safely stored in the Deutsches Museum's vaults in Munich, Germany (Fussl, 2010). During April-September 2003, the Deutsches Museum exhibited "Klima: das Experiment mit dem Planeten Erde" ["Climate: the experiment with the Earth"] and some display space was allotted to Herman Sorgel's ATLANTROPA PROJECT. The 1929-1952 AD planning effort towards materialization of the ATLANTROPA PROJECT always had, as its key infrastructure, the super-massive concrete Strait of Gibraltar Dam. [Circa 1929 AD was about the time the word "space suit" first appeared in English-language dictionaries.] **Figure 6a**, below, offers a memorable illustration of the materials requirement in terms of sheer bulk, by juxtaposing Mars's natural satellite Phobos with Gibraltar Strait succeeded by a sketchily blueprint-like version of the monumental hydro-electric facility, **Figure 6b**, designed and macro-planned mostly by Herman Sorgel himself.

Figure 6a. Phobos incongruously inserted into the North Atlantic Ocean-Mediterranean Sea physical geographical gap. Team Geographos approached a consensus opinion that Phobos, when plunked down into Gibraltar Strait in this proprietary Team Geographos image, makes an impression on the astute reader that is much the same as that gotten from **Figure 9** in Chapter One.

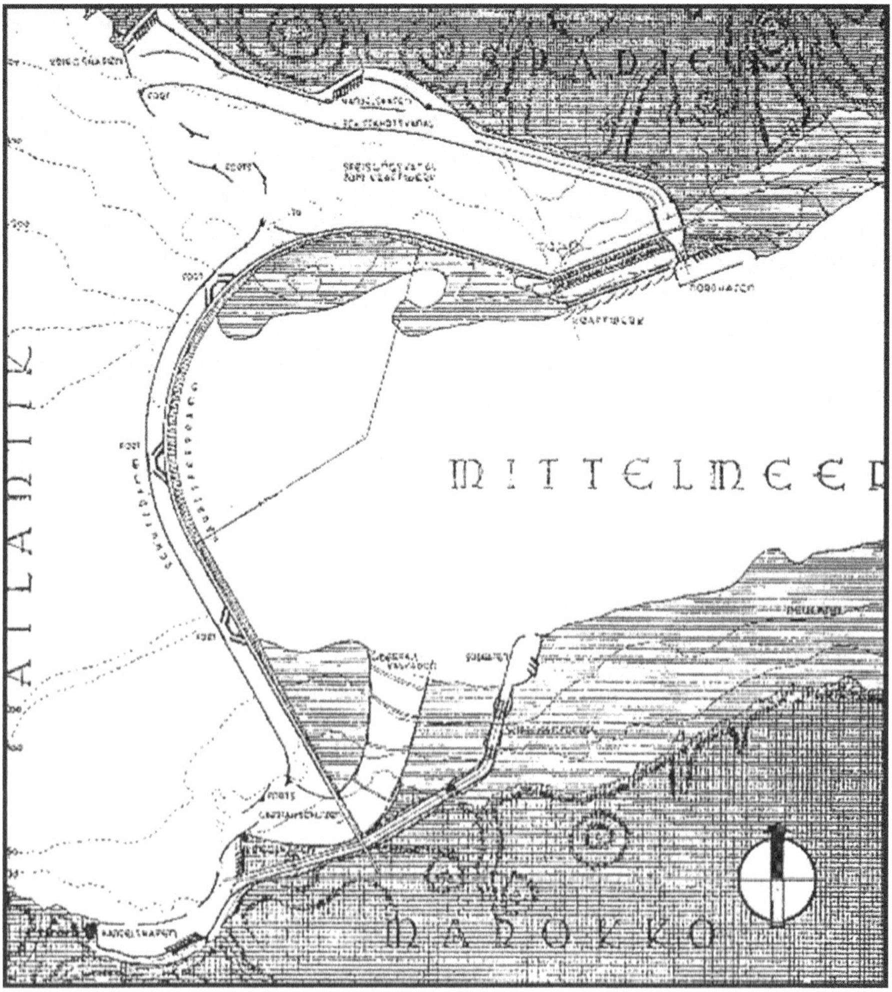

Figure 6b. ALANTROPA PROJECT's gargantuan Gibraltar Strait Dam, shown in this "blue-print" planning document with all of its 1932-version appurtenances and frills, which include outlying fortresses, actual "fore-guards" like the North Sea's Heligoland (Ruger, 2017) in warfare, not avant-garde, and small naval defense vessel ports, strung anti-torpedo nettings, as well as emplaced seawater-current deflectors and blockading impediments! (Image courtesy JMH.)

According to the third chapter of ***Star Trek: The Motion Picture*** (1979), by Gene Roddenberry (1921-1991), during the 23rd Century [circa Star Date 7412.3], Commander James Tiberius Kirk visits a massive old hydro-electric complex at Gibraltar just before embarking on a vital Starship mission. There follows a clear-cut tourist guide's description of Herman Sorgel's ATLANTROPA PROJECT, one of the very few references to be found in either popularized science, or in that literature genre commonly known as science-fiction [Sci-Fi]

(Cathcart, 1983, page 120-121). An Internet search netted "No Mediterranean Sea/Alternative History/Fandom powered by Wikia" [GOTO: http://althistory.wikia.com/wiki/No_Mediterranean_Sea] which offered a pallid, uninspired and nerdy alternative history spin-off, without attribution, of the classic 1999 Sci-Fi novella ***Down in the Bottomlands (and other places)*** by Harry Norman Turtledove (born 1949) and L. Sprague De Camp (1907-2000). Set and transpiring 5.5 million years in our Earth's past, the Turtledove-De Camp story features Homo neanderthalensis characters inhabiting a unique deep Earth-crust trough hot dryland region (Schneck et al., 2010), Trench Park, as storyline stand-in for the Mediterranean Sea of 5.5 million years ago, protected from world-ocean seawater inundation by the blocking Barrier Mountains. Dictatorial villains greedily seek to vaporize and/or otherwise remove the Barrier Mountains (causing the flooding of Trench Park) with a sufficiently high-yield secretly planted thermonuclear bomb explosive; in other words, the Turtledove-De Camp storyline twist invalidates the commonplace anthropological assumption that only Homo sapiens can ever produce a tool—no less than an fission-fusion "atomic" bomb in their fascinating Sci-Fi novella—for a specific future purpose. Team Geographos wonders if the Turtledove-De Camp team recognized the similarity of their science-fiction to the historical real-world geophysical and anthropogenic process-events that permitted the Salton Sea to accumulate in the Salton Trough early in the 20th Century. (Conspicuously, the villain-architect's evilly destructive planned asymmetric warfare strategic surprise, rapid submergence by an irresistible flood of turbulent seawater spilled into Trench Park from the Atlantic Ocean, is a failure.)

21st Century Mindset Macro-projects: Our Greater Mediterranean Sea Basin

Before elucidating Team Geographos' views on feasible large-scale anthropogenic ceno-Anthropocene reconstruction of the Basin's sea-adjacent landscapes, it is really necessary to examine all the contributory landscapes that geographically ring the Mediterranean Sea Basin. ***World Macro-Imagineering: Masterpiece Mega Projects*** (2017) was mostly centered on possible macro-projects that may someday alter the ecosystem-nations bordering the Indian Ocean; so, this book is an extension of our thoughts northwestwards to the "Middle East", the "Mediterranean Sea", "northern Africa" and "Europe" (Morton, 2017). **Figure 7,** below (Google Image).

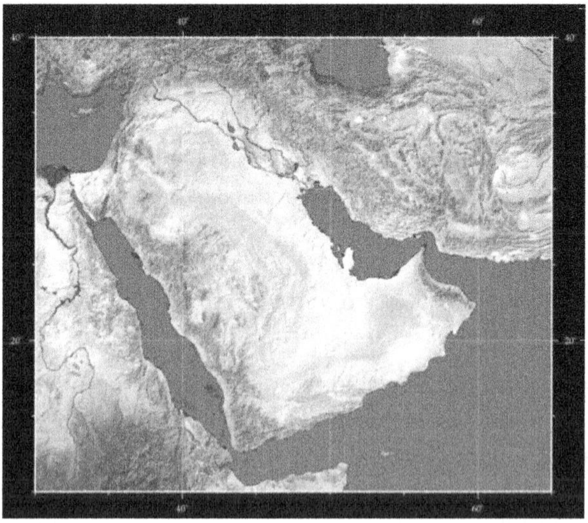

In **Figure** 7, Team Geographos sought to highlight the geographical positions of the Arabo-Persian Gulf [mid-level, right-side of framed map] and the Red Sea [the left-side of map] because these bodies of seawater may, during the remainder of the 21st Century, become vastly modified by human-designed and emplaced constructions of enormous bulk that is not much less than Herman Sorgel's imagined impermeable monumental concrete edifice for Gibraltar Strait. As a semi-enclosed marine system, the Arabo-Persian Gulf's saltwater, air-quality, and climate are undergoing substantial changes caused, in part, by an array of exclusively anthropogenic actions and infrastructures. For example, there are more than 800 offshore oil and natural gas platforms; effluent brine—typically two cubic meters of waste brine is produced for every cubic meter of freshwater—issues from that region's many huge seawater desalination plants (Smith et al., 2007); enormous volumes of granular material have been taken from the seafloor, moved and been deposited by dredgers to make artificial island-lagoons built close to the national territorial shoreline (Naser, 2015; Gupta, 2015; Elhacham and Alpert, 2016), and hydrocarbon air pollution (McLinden et al., 2016) still spews from several dozen oil terminals as well as the diesel fume-emitting funnels of 25,000-50,000 tanker ships plying its waters annually. At least 800 offshore production sites dot the seascape. Saudi Arabia has announced plans for the world's first large-scale solar-powered desalination facility to be located in Al Khafji, near Kuwait (Sarant, 2016). Since the likelihood is definitive that the Arabo-Persian Gulf's climate regime will become warmer than nowadays, desalination plants are probably going to be erected in many other places near the coastline, possibly even introducing tropical cyclones which have never before been recorded by modern literate historians (Lin and Emanuel, 2016)! Succinctly, one group of experts, not "Scienceers", unequivocally have stated: "The [Arabo-Persian] Gulf: A young sea in decline" (Sheppard et al., 2010, page 13). Because the technological and sociological modernizing of the region's human population is ongoing, and the populace is becoming more numerous and becoming more urbanized as well, fixed point-sources of landscape and air-mass changes are gaining in eco-social influence. From 3300-2600 BC sea-level in the Arabo-Persian Gulf stood one-meter higher than during AD 2017 before it reduced to early-21st Century levels by about the time the Roman Empire was at is greatest extent geographically [117 AD] (Lokier et al., 2015). So, any future regional seawater level rise of one-meter is, in effect, a duplication of long-forgotten geophysical process-events indisputably operative 5,300 to 4,600 years ago affecting 21st Century coastal landscape!

The Arabo-Persian Gulf is about 1,000 kilometers in length and ranges from 350 kilometer maximum width to only 5.0 kilometers wide at the Strait of Hormuz. The Arabo-Persian Gulf is the shallowest major inland extension of the world-ocean with a mean depth of less than 50 meters. During the most recent Ice Age, Earth's ocean sea-level was markedly lower than the 21st Century's current, and gently rising, sea-level. Thousands of years ago, during the last glacial maximum period, people lived continuously in a valley that is called humanity's current Arabo-Persian Gulf (Kennett and Kennett, 2006). As the post-Ice Age period commenced, the world-ocean penetrated and permanently submerged the valley—it is possible a voluminous barrier of previously wind-deposited sand piles blocked the Strait of Hormuz until the dune-field was washed away, perhaps quite suddenly, by torrential rainfall caused by intruding tropical cyclones and by rapid erosion from powerful impacting sea-waves east of the hypothesized sand deposit (Teller et al., 2000). During 2005 AD, macro-engineers proposed emplacement of a solid concrete equivalent of the hypothesized granular dune-field connecting Saudi Arabia and Iran with a hydro-power plant at the Strait of Hormuz the physically separates the Arabo-Persian Gulf's seawater and life-form content from the world-ocean's (Schuiling et al., 2005). Any future regional rise of sea-level will naturally affect the Shatt al-Arab River, which originates from the confluence of the Euphrates and Tigris rivers, by pushing saltwater upstream from the mouth: during a future very-low river runoff period, seawater could intrude more than 100 kilometers upstream (Abdullah et al., 2016). Low freshwater flow at the Shatt al-Arab's mouth may be caused by meteorological drought and by human upstream reservoir users as well. Obviously, an increased domestic freshwater supply is a pressing need of settled people in eight ecosystem-nations, totaling 168 million persons, living within the Arabo-Persian Gulf's coastal strip region. Team Geographos Sciencers may be helpful!

Deploying New, BIG Technologies in the Arabo-Persian Gulf

The spatial extent of hypersaline volumes of water is limited; brine accumulations such as the inadvertently artificial lake below sea-level Salton Sea are uncommon. The Arabo-Persian Gulf seawater is characterized by elevated salinity and warm temperatures due to its shallowness, strait-restricted water exchange with the world-ocean and its 1-2 meter/year equivalent of evaporation. The western shoreline of the Arabo-Persian Gulf has numerous enclosed bays which are very shallow, have limited seawater flushing action and equally high evaporation rates—the bay which has the greatest salinity and temperature variations in the western Araba-Persian Gulf is the Gulf of Salwa, **Figure 8**, because it is nearly isolated by the island of Bahrain

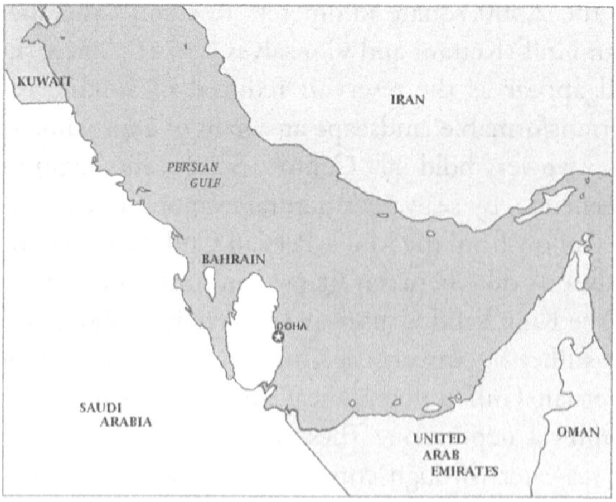

(Mugnier, 2007) and the two-lane 25 kilometer-long King Fahd Causeway situated since 1985 between Saudi Arabia with Bahrain. [The Gulf of Salwa is a primary conservation region for the endangered dugong [Dugong dugon] which grazes seagrass growing therein (Loughland, 2012).]

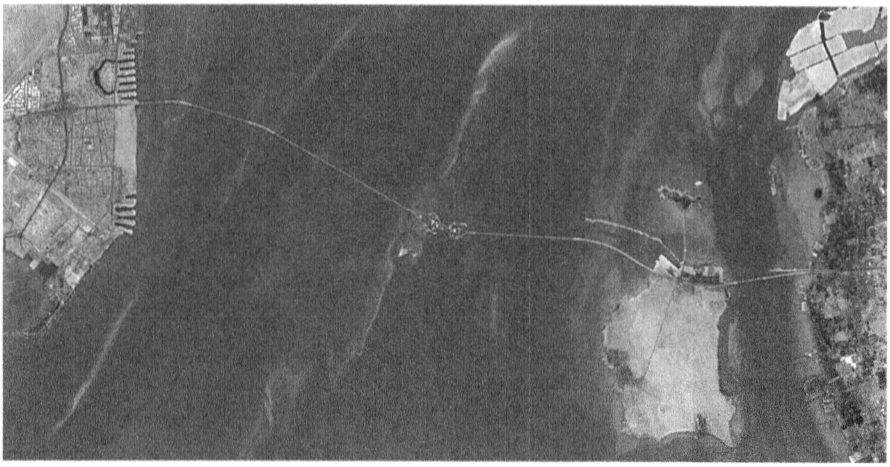

The King Fahd Causeway, satellite-pictured above in Google Image **Figure 9**, is paralleled by the A8 active oil pipeline carrying fluid extracted from the Bahrain Oil field; Qatar's west coast-sited Dukhan Oil Field shifts petroleum fluid by pipeline and natural gas by two pipelines to its east coast seaport of Umm Said. Located at latitudes 24^0 40' to 25^0 50' North and longitudes 50^0 10' to 50^0 50' East, the 110 kilometer-long funnel-shaped Gulf of Salwa is 45 kilometers wide at its entrance/exit. Maximum seawater depth of the conical Gulf of Salwa is about 17 meters nearest Saudi Arabia. During 1970, Dr. M. Ali Kettani (born 1941) and Lenine M. Gonsalves proposed a mega-project to impose a useful "land" barrier between the 2,300 square kilometer "reservoir" and the vaster remaining Arabo-Persian Gulf (Kettani and Gonsalves, 1974). A new shoreline—sunny beaches!—would appear as the reservoir reduced in volume to optimum size, amounting to a transformable landscape area gain of approximately 4,100 square kilometers. These two very bold 20th Century Sciencers sought to use impinging solar energy to generate, by seawater's natural evaporation, a practically constant artificial seawater inflow from the Araba-Persian Gulf into a geographically diminished Gulf of Salwa. A dike between Qatar Peninsula and Bahrain, along with a modification of the King Fahd Causeway to make it a solid dam-like water body separator, would suffice to convert the Gulf of Salwa into a briny reservoir from a mere Arabo-Persian Gulf topographical appendage—that is, into an anthropogenic topographical depression. These dike-barriers, penetrated by conduits feeding flowing seawater through corrosion-resistant hydraulic turbines, would have an optimum seawater drop of 12.4 meters from the Arabo-Persian Gulf into the reduced Gulf of Salwa. Their facility, allowing the generation of approximately 300,000,000 kilowatt-hours of electricity, would cause only about a 0.25 millimeter complementary rise of the extra-Gulf of Salwa world-ocean sea-level; if all our world's ships were removed from Earth's ocean, the world-ocean would shrink in volume, lowered by about 0.0005 millimeters!

Large Horizontal Dimension Deformable Structure: Beyond Mere Barges!

The 19th Century science-fiction novelist Jules Verne was probably the first to write about areally large artificial floating islands in ***L'ile a helice*** (1895)—that is, a deformable structure, such as Japan's Mega-Float, that is significantly bigger than the largest ship afloat (Gerostathis et al., 2016). The first large, mobile floating aircraft runway, Project SOCK ["Special Operations Capable Key"—"key", as in a State of Florida "island"] was constructed by the U.S. Navy Civil Engineering Corps during 1943 for wartime testing purposes as a means of military access, control or possession. **Figure 10**, below.

Long enough to handle World War II propeller-driven fighter and patrol aircraft take-offs and landings, Project SOCK was composed of linked standard T-6 steel box pontoons, with tapered T-7 steel pontoons attached to the ends [bow and stern, depending on direction of movement]. It was intended to navigate the open waters of the world-ocean. Segmented steel pontoons were first designed and used for limited purposes 1935-1937 AD. The T-6 box pontoon displaced 4.955 cubic meters and floated in just 0.457 meters of seawater, a single pontoon could support about 4.989 metric tons through buoyancy. [During 2016, the U.S. Navy acquired and deployed its Improved Navy Lighterage System, a sectional craft with floating sections that lock together to create a variety of floating structures such as causeways, docks and ferries. These INLS craft allow U.S. Navy transports to offload and efficiently shuttle heavy-equipment between offshore supply ships and shore bases, when moorings are damaged or unavailable. The Improved Navy Lighterage System can operate in Sea State 3 and survive in Sea State 5.]

The Strait of Hormuz, just a few tens of kilometers wide, is a true bottleneck—that is, its geographical existence as the Arabo-Persian Gulf's sole entrance/exit is a localized navigational constriction on ship passages, a limitation in the voyages of maritime shipping traffic with obvious international geopolitical import (Alexander, 1992). However, most of the ecosystem-countries of the region are gradually weening themselves from a dwindling petroleum-natural gas resource base as well as constructing long-distance pipelines to transport their exported oil and natural gas, trying in some instances to deliberately avoid the Hormuz Strait navigational bottleneck. Some macro-imagineers have offered at least two bypassing canal options. **Figure 11a**, below (Google Image).

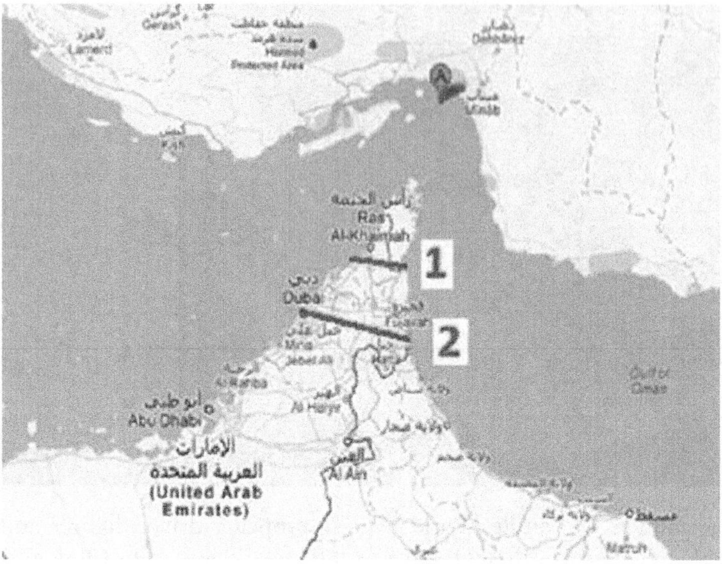

There is a slight possibility that one of these canals, or a pipeline following a similar route, could be used to deliver freshwater supplies derived from icebergs harvest from Antarctic, if the project, said to commence in early-2018 AD is fully implemented by the National Advisor Bureau Limited [www.nationaadvisor.ae], headquartered in Masdar City, Abu Dubai and presently headed by Abdullah Mohammed Sulaiman Al Shehi. **Figure 11b**, below shows the extraction shipping route and its appointed destination, the Fujairah region of the United Arab Emirates.

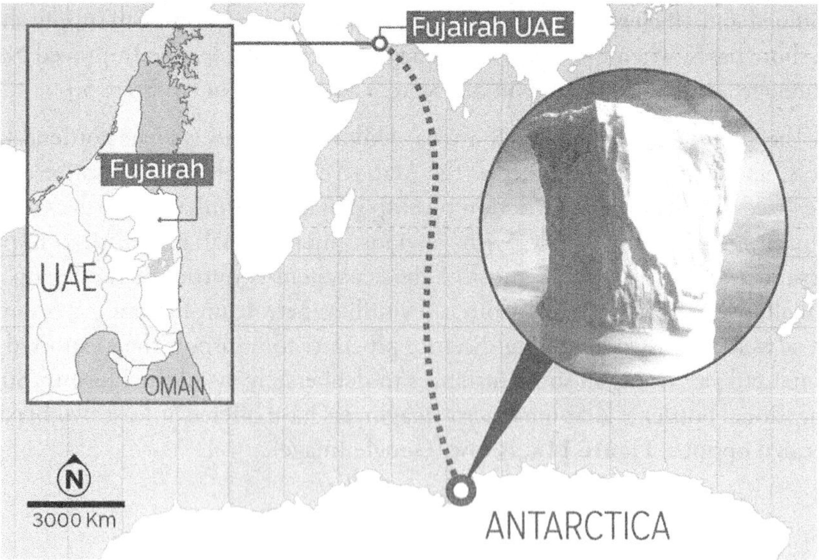

Dry Canals have been proposed for the Kra Peninsula in Thailand by the Korea Railroad Research Institute (established 1995); France once considered a 1939 macro-project plan to carry ships 483 kilometers overland by six-track railway to link the Mediterranean Sea with the North Atlantic Ocean (Port le Verdon to Beziers). Ship-designers, freight carrier companies and maritime traffic-coordinators already expect that by, say, AD 2030, remotely-controlled unmanned cargo ships will be operational on the world-ocean and that by approximately five years later [circa AD 2035] fully autonomous cargo ships will be plying the world-ocean. It seems hardy, sometimes lonely seafarers are a vanishing employee group! Furthermore, new plans for amphibious vehicles such as sturdy improved very large hovercraft—invented by Christopher Cockerell (1910-1999)—will likely produce a new transport economy in the Arabo-Persian Gulf region because such vehicles can move over seawater and flat hot roadless desert landscapes easily. It is foreseeable that floating dry-docks will be required to service remote-controlled and autonomous maritime transports as well as sections of our proposed CTERB described later. A dry-dock is a vessel that can be partly submerged by seawater flooding of ballast tanks, then refloated by pumping out the tanks once a repairable or serviceable ship has been floated inside the dry-dock and safely settled on keel-blocks and side-blocks, in order to perform the required maintenance and repairs that could not be accomplished whilst the ship in need was still afloat outside the dry-dock. Since the probability has increased that tropical cyclones will enter the Arabo-Persian Gulf in future, both ships and dry-docks must be capable of withstanding the sweeping wind and pounding waves of a tropical cyclone as well as major earthquakes possibly caused by previous and ongoing natural gas, oil and groundwater resource mining operations. In two previous books—***World Macro-Imagineering: Masterpiece Mega Projects*** [Chapter Seven, pages 158-163, 2016] and ***Sci-Fi Macro-Imagineering*** [Chapter Three, Figure 3, 2017]—the "Cathcart Thermal Energy Rejection Barge" (CTERB), an ultra-large scale flexible floating structure, was thoroughly described in both its constitution and contributions to a healthy environment within the Arabo-Persian Gulf. CTERB, like a region-wide power and air-conditioning system, cannot ever fail since heat-waves in the region that produce wet-bulb temperatures of 32^0 Celsius, or higher, are human life-threatening! Since the human body operates at 37^0 Celsius, its metabolic efficiency in an aerial environment of 37^0 Celsius would be zero—that is, the total energy of all food consumed would be wasted and human life could not be sustained! Above 37^0 Celsius the human body sweats profusely to maintain life-function and that has to be compensated by evaporation from the skin and lungs which requires 0.18 liter per hour of freshwater to be drunk—putting the onus on macro-engineers operating desalination plants in a freshwater-short hot desert region such as the Arabo-Persian Gulf. However, Team Geographos now offers a few further refining remarks thereon. The proposed CTERB, a modularized linked set of barges, is truly gigantic as it would be hundreds of kilometers in length and width, with a shallow draft and only a few aerial projections that might offer obstruction to low-flying aircraft or leverage points for wind. SEE: **Figure 12**, below (RBC Image), an exaggerated sizing since the aerial Physics remain unknown. Fabricated with slabs of magnesium alloy matrix composite so light-weight (because it is a syntactic metal foam) that it floats in seawater without any vessel compartmentalization (Anatharaman et al., 2015); such a gigantic barge-like CTERB ought to be "unsinkable", although Team Geographos is loath to use that post-April 1912 **RMS Titanic**-haunted term! Any future existence of the CTERB moored or anchored within the Arabo-Persian Gulf might become of interest artists and advertisers—would it truly be uglier than other forms of well-known Art such as blimps equipped with blinking lights, long banners towed by slow-moving airplanes or smoky-trail skywriting? Team Geographos' main interest, though, is the favorable postulated beneficial affect an exaggeratedly gigantic CTERB should have on the climate regime of the Arabo-Persian Gulf. Its feasibility is utterly dependent on Physics of both air and seawater as yet unexamined by experts—that is, those persons who have a recognized scientific and engineering career and are, therefore, possessed of some immunity to uninformed or outrageous public opinion. CTERB Macro-project does not preclude either the Hormuz Strait Dam Macro-project or the Gulf of Salwa Macro-project, nor interfere with any continuing oil and gas extraction efforts or tanked transport.

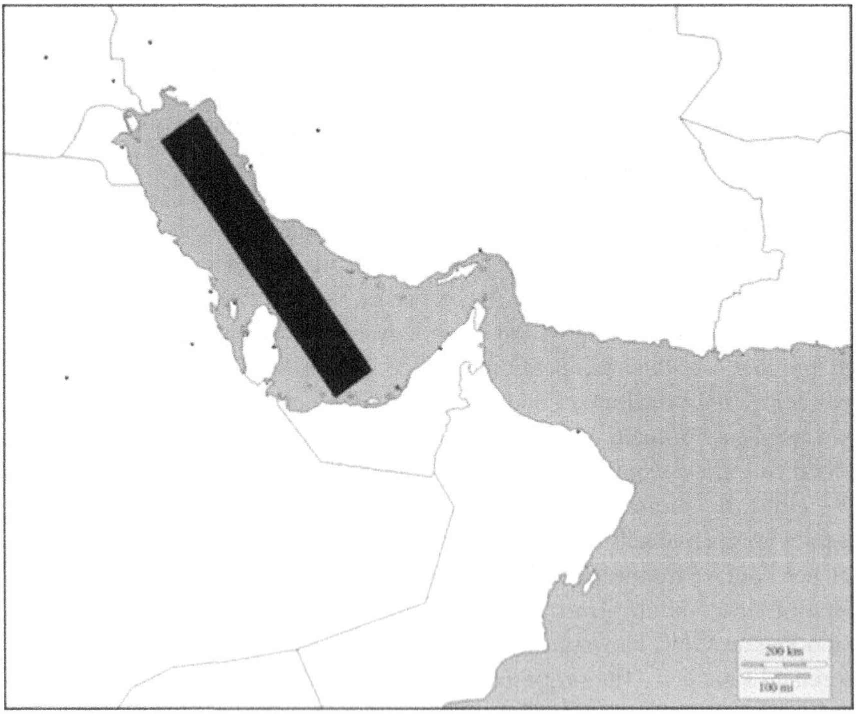

A new, cheaply bubble-wrapped "Solar Vapor Generator", which floats stably on seawater, without costly optical sunlight-concentration spectrally-selective absorbers, has been demonstrated to heat seawater, drawn from below the "Solar Vapor Generator", to its boiling point of 102^{0} Celsius, thus converting about 20% of the impinging solar energy into steam (Ni et al., 2016). In other words, every "Solar Vapor Generator" can boil seawater to inexpensively produce steam under ambient solar flux! [Bubble-wrap was invented in 1957 by Alfred W. Fielding (1917-1994) and Marc A. Chavannes (1896-1994) as a Space Age plastic wallpaper product that was commercially unsuccessful. Later, it was re-invented by Fielding and Chavannes to become eventually the cellular air cushioning used by the packaging industry and manufactured by Sealed Air Corporation (established 1960).] Any useful application, and appropriate structural installation, of multiple Solar Vapor Generators to the spacious uppermost deck of the suggested Arabo-Persian Gulf "Cathcart Thermal Energy Rejection Barge" [CTERB] would likely produce a dense localized air-mass enriched with freshwater vapor above each emitting field of arrayed Solar Vapor Generators.

Water (fog, dew and vapor) of the lower Earth-atmosphere has been an extracted resource done by artificial means for much of human history (Klemm et al., 2012; Khalil et al., 2016). Some macro-engineers, specializing in water-supply macro-problems conduct R&D related to better ways to promote and transport condensed water droplets. The major challenges in harvesting freshwater from air are controlling droplet bulk as well as the speed in which they form and the direction in which they subsequently flow. The formation of water droplets by use of a widely-deployed field of Solar Vapor Generators provides an industrial means for the creation of multiple captive and controlled miniature anthropogenic rainstorms from desalinated seawater in regions of suitable dry climate regime such as the Arabo-Persian Gulf region (Starr and Anati, 1971; Starr et al., 1974; Forootan et al., 2017)! In brief, Homo sapiens may soon discover the full consequences of actual global freshwater management practices beyond the unnatural terrestrial component of our Earth's natural hydrologic cycle induced by our ever-expanding networks of ecosystem-nation reservoirs, canals, pipelines, groundwater removal mega-projects, reservoir coverings of floating sunlight reflectors to limit evaporation and, of course, efficient commercial-oriented crop channel and drip irrigation macro-schemes.

Basically, and briefly, Victor P. Starr (1909-1976) and David A. Anati along with all their eventual team, colleagues postulated an light-weight artificial vertical tube with a height of three kilometers and a radius of fifty meters which they named the "Aerological Accelerator". **Figure 12**, below, taken from "The Control of the Water Cycle". *Scientific American* 228: 57 (April 1973) written by Jose Pinto Peixoto (1922-1996) and Ali M. Kettani (born 1941). [SEE also: Alexander Alexandrovich Bolonkin (2011) "Production of Freshwater and Energy from Earth's Atmosphere", *Smart Grid and Renewable Energy* 2: 86-98.] Simply stated, warm moist air rises until it chills enough to form clouds which can then be harvested of their freshwater droplets by using appropriate mesh webbings or even magnetic means.

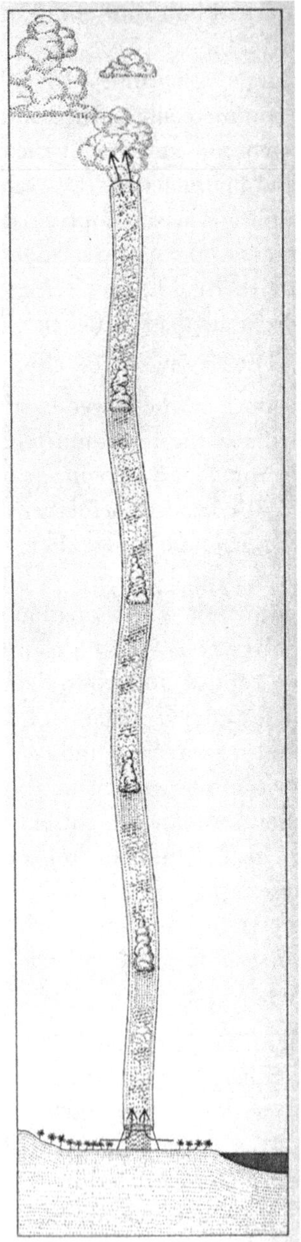

The Starr and Anati "Aerological Accelerator" is a twin-layered vertical tube comprised of gas-impermeable film that is buoyed by methane, a main component of natural gas. With a specific density of 0.72 kilograms per cubic meter, methane provides the lifting force in air, which has a specific density of about 1.225 kilograms per cubic meter. In other words, every cubic meter of methane can hold aloft a 0.5 kilogram payload. Each "Aerological Accelerator" may be stabilized vertically and horizontally by nanotube super-ropes as well as drones. The "Aerological Accelerator" enables a major export of water vapor downwind and, therefore, affects the far-field climate regime. Such a device works by lifting warm surface air laden with evaporated moisture much higher in the atmosphere than the lifting condensation level of freshwater—that is, where the humidity of the mass of lifted air reaches 100% when it is cooled by aerial adiabatic lifting. Thus, freshwater may be mined from hot moist air parcels as condensation collected by a fine mesh fabric (Park et al., 2013) installed at, or slightly below, the lifting condensation level and the droplets accumulated are dropped to earth gravitationally by sliding, duplicating Nature's atmospheric convection process-event (Ming et al., 2017). Summarizing, an Arabo-Persian Gulf enhanced "Cathcart Thermal Energy Rejection Barge" (CTERB) will be multi-use "rural" infrastructure: radiating heat away from Earth's bioshell to outer space for planetary cooling, providing a floating, yet mostly immobile base for copious anthropogenic water-vapor generation and, finally, a secure foundation for many physically awesome projective "Aerological Accelerators". In other words, a "landscape" of active infrastructure situated atop a shaded seascape with installed entities composed partly of natural systems and partly of machinic systems! Because the facility—its barge-like base vessel support as well as all of its superstructure protuberances—surely will become the region's iconic symbolizing and actualizing form of Arabo-Persian Gulf human population freedom, economic prosperity, emblematic of ostentatious wealth and, thus, it certainly needs a proper, memorable moniker. Since the word "utopia" really designates another referent—the "other" of any particular region—how about "Cyborg Arabo-Persian Gulf Utopia" (CAPGU)? Pronounceable, CAPGU will fit nicely with the plethora of bothersome acronyms listed in today's irritatingly thick guides to scientific and engineering abbreviations!

The Red Sea Dam Macro-project, or the Dead Sea's Rejuvenation

With an Earth-ocean intake at one elevation and a Red Sea outlet at some lower elevation, the potential hydroelectric power generated by hydraulic turbines from such a dam-caused powerhead at the 29 kilometer-wide Bab-el-Mandab Strait is proportional to the rate-of-seawater-inflow and to the difference in elevation of the two places of a single massive concrete dam (Hassan, 1961; Schuiling et al., 2007). The Red Sea separates the Arabia tectonic plate from the Africa tectonic plate—thus it is a geographical Earth-crust indentation that is a longitudinally elongated depression in prolonged transition from crust-rifting to crust-shifting, a mark of active seafloor spreading (Almalki et al., 2015; Augustin et al., 2016). Nearly 2,000 kilometers long, with a maximum width of about 280 kilometers, the 21st Century Red Sea extends from latitude 12.5^0 to 30^0 North; at the north end, the Red Sea branches into the subsidiary Gulf of Suez and the Gulf of Aqaba. [The shallowest part of the Red Sea's southernmost segment is Hanish Sill, 137 meters deep, in the vicinity of the Hanish Islands almost 100 kilometers north of the Bab-el-Mandab Strait (Werner and Lang, 1975).] Nevertheless, any Red Sea Dam would impose, wherever it was ultimately emplaced, more than the Bab-el-Mandab Strait already, a hydraulic valve-like regulation of world-ocean seawater inflow to Nature's strongly evaporative Red Sea. Done strictly as an extrapolative Team Geographos gedanken-experiment, the Red Sea Dam Macroproject is infeasible on account of long-term structural safety macro-problems as well as purely economic reasons and much less worthwhile than the Team's Hormuz Strait Dam Macroproject. [The Red Sea is the only rift region of our world defined by two parallel very long escarpments which separate high-plateaus from narrow coastal plains.] However, a 28 kilometer-long Yemen-Djibouti Fixed Link—a record-setting suspension bridge mostly likely—was proposed during AD 2008 that would utilize Perim Island. **Figure 13**, shows crossing's suggested approximate superstructure alignment (Google Image).

During the Ice Age, humans transited the seawater-gap separating Africa from Arabia without the use of seaworthy boats or experience-acquired seafaring skills—it was then a clearly-visible narrow, perhaps less than five kilometers—quite an inviting adventure to our wandering ancient ancestors (Lambeck et al., 2011; Gandini et al., 2016). Post-Ice Age times of clement regional climate regime prompted people to relinquish nomadism as a life-style and to settle near the Bab-el-Mandab Strait as well as at other crossable beach-strand and wider coastal places of the Red Sea's shoreline (Reba et al., 2016) where advantage could be taken of the sea's whole ecosystem and discernable fisheries (Tesfamichael and Pauly, 2016). Since the 1970s, moderns have investigated the possibilities of mining of valuable metalliferous muds sitting in the Red Sea's depths (Jones et al., 2017). Team Geographos members have devised a feasible and economical means to dispose of anthropogenic brine waste pumped from desalination factories prevalent in the region into the greatest depths of the Red Sea where it should pool by density and gravity for "eternity", finally out of 21st Century human sight and probably to be overlooked, even forgotten, by many distracted human minds (Badescu et al., 2013). If climate regime change causes a one-meter eustatic sea-level rise of the Red Sea, then Saudi Arabia will lose about 890 square kilometers of landscape from its 1,840 kilometer-long strand whilst Egypt might lose 106 square kilometers of its Red Sea coastline. Were the misbegotten Team Geographos' Red Sea Dam Macro-project undertaken, it would make impossible just about any imaginable Dead Sea strategic revival megaproject. It is notable that a one-meter rise by AD 2100 would result in a cumulative load pressure on the world-ocean's floor of 10 kilopascals; super-computer modelling studies seem to indicate that any major vertical sea-level change, about 100 meters either added or subtracted, are capable of triggering and/or modulating tectonic activity directly by Earth-crust flexure and subsequent seismic energy releases (Brothers et al., 2011). So, there might be some repercussions seismically to the Dead Sea's rapid enlargement and consequent Earth-crust loading.

Djibouti, once an almost forgotten mini-state ecosystem is now in the forefront of international relationships essentially because it has become a seaport and air-shipment hub meant to serve China's 21st Century Maritime Silk Road (Yang et al., 2016). A modern railway from Djibouti to Ethiopia, financed by China, opened in AD 2017. Ethiopia has completed its Grand Ethiopian Renaissance Dam regulating the Blue Nile River; when fully functional, it will soon generate 6,000 Megawatts of electricity which earns foreign currency from Egypt and Sudan (Taye et al., 2016). Previously, Team Geographos proposed a novel geothermal resource exploitation scheme focused entirely on Djibouti's famed below sea-level Lac Asal (Rugescu et al., 2011); were Lac Asal connected to the close-by ocean by an excavated ocean-shipping canal, it would prove to be a nicely wind-sheltered seaport, possibly even a small inclement weather-secure almost landlocked future China naval base! Such expansion of the Autonomous Port of Djibouti takes full advantage of the most capable rail-and-seaport infrastructure on Africa's east coast.

Dead Sea Freshwater Production Macro-project

Bitter Waters: America's Forgotten Naval Mission to the Dead Sea (2011) by David Howard Bain virtually ignored the Dead Sea's geological formation, instead focusing on the 1847 AD explorations laboriously carried to conclusion by Lieutenant William Francis Lynch 1801-1865). Team Geographos must pursue the same course of action for the straight-forward reason that the Dead Sea's actual geological formation is still incompletely investigated geologically or fully understood by geoscience experts. Instead, we will cope only with the Dead Sea's unique geography. Long-ago lake water levels have been estimated by measurements taken from bored lake-bottom halite sediments (Kiro et al., 2017). [If all salt were extracted from the world-ocean and spread evenly atop Earth's land it would form a stratum of salt approximately 166 meters thick.] Long ago, the Red Sea, Dead Sea and Mediterranean Sea were connected through the Dead Sea Fault Zone (Meric et al., 2016) Today's Dead Sea ponded water is almost ten times saltier than the world-ocean's water (Katz and Starinsky, 2009). Any person frolicking in the briny waters of the Dead Sea takes a big personal risk of drowning or being poisoned, mainly because to accidently ingest a quantity of the highly salty waters therein is to quickly induce catastrophic kidney failure (Yagil et al., 1985). The Dead Sea crust depression feature has two potential landscape water overflow outlets were it filled to global sea-level artificially: (I) 60.5 meters elevation into the Mediterranean Sea via the Jezreel Valley, near the town of Afula, and another (II) 200 meter elevation into the Red Sea via the Arava Valley; the maximum area of (I) produces an enlarged Dead Sea with and area of 5,985 square kilometers and a lake volume of 1,602 cubic kilometers (Hall, 1996). **Figure 14**, below (Google Image) illustrates an enlarged Dead Sea at Earth's extant "global" sea-level. Is it time to rethink "The Negev Continental Bridge" (Stern and Gradus, 1983), perhaps by redirection of overland traffic from Eilat on the Gulf of Aqaba toward the enlarged Dead Sea instead of to Ashdod or Haifa seaports?

Whilst there are extant many proposed macro-engineering plans for refreshing the Dead Sea by raising its AD 2017 water level through the importation of seawater extracted from the Red Sea's eastward extension named the Gulf of Aqaba and from the Mediterranean Sea (Malkawi and Tsur, 2016; Aggestam and Sundell, 2016; Al-Omari et al., 2014; Bashitialshaaer and Persson, 2011; Malkawi et al., 2010; Qdais, 2008; Beyth, 2007), Team Geographos prefers to apply Macro-Imagineering's own "Occam Razor"—explanatory simplicity of a complex infrastructural mega-project conception—to the fascinating macro-problem of the Dead Sea's aquatic future. For instance, the New York architect Phu Hoang Office, LLC [http://www.phuhoang.com], established 2006, during 2007 offered its peace-promoting floating islands mega-project ***No Man's Land***, which was to be centrally sited on the present-day Dead Sea shared by Israel and Jordan, the ultimate product of which was to be freshwater harvested from the air that could then be exported to nearby regions suffering freshwater stress or actually shortages! In other words, Phu Hoang Office, LLC presented an early version of Team Geographos' own CAPGU ("Cyborg Arabo-Persian Gulf Utopia"). Mr. Phu Hoang, **Figure 16**, teaches advanced design studios and seminars at Columbia GSAPP in New York City, State of New York. Unless an impervious six kilometer-long barrier—such as a solid causeway rather than a bridge that did not hinder shipping traffic or merely an heavy-duty aerial cable-propelled transit and freight system using hung gondolas—was emplaced at the Strait of Tiran linking Saudia Arabia-Tiran Island-Egypt the seawater flow into the Gulf of Aqaba could not be interrupted. Saudi Arabia and Egypt agreed during April 2016 to begin to build a Strait of Tiran fixed link between their currently allied ecosystem-countries soon. **Figure 15**, below (Google Image).

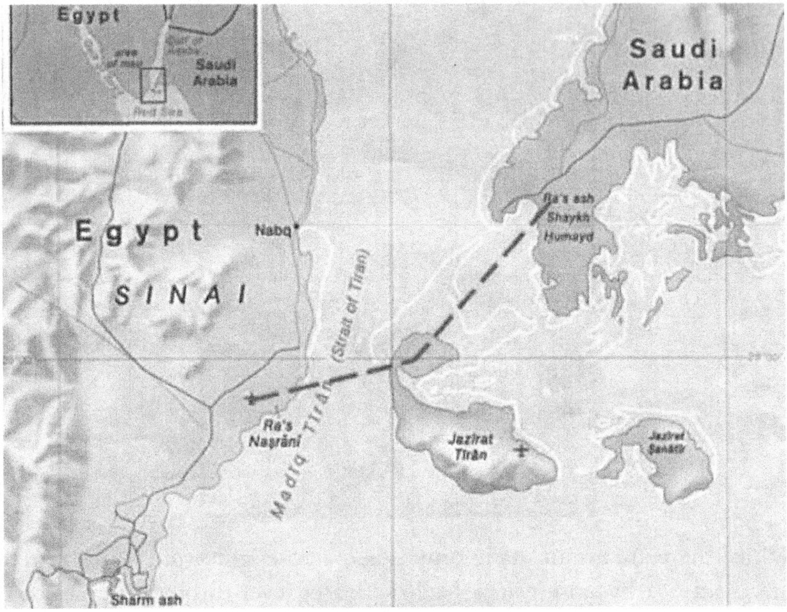

Figure 16. Mr. Phu Hoang, above, world-renowned architect of ***No Man's Land***, pictured and detailed in Mark Fletcher's magnificent opus ***Islands: Contemporary Architecture on Water***. (H.F. Ullmann (2010) at pages 342-345.

Enlargement of the Dead Sea, Earth's deepest continental landmass depression, to a physical size where it is equal to today's "global" sea-level—that is, a new body of seawater-equivalent about 250 kilometers long—which varies regionally due to many factors, must of necessity lead to the abandonment of operational hotels, a unique far below-sea-level military air-field sometimes used by some private and commercial tourist flights, many kilometers of circuitous paved roads, furnished beach-parks and possibly even cosmetic manufactories (Abdel-Fattah and Pingitore, 2009; Charlier and Chaineux, 2009). But, the more voluminous, diluted induced Dead Sea will provide a more or less permanent, at least practicably sustainable by hard-working future macro-engineers, seawater-defined, de-militarized neutral zone linking Jordan and Israel, a resource for aerial humidity harvesting of great socio-economic value to both ecosystem-states, and even the real possibility of increased downwind natural precipitation on desert landscapes of both Jordan and Israel, caused by evaporation over a huge anthropogenic area of Sun-exposed seawater, as well as downwind cloud formations that may be anthropogenically seeded with water droplet-inducing nuclei to produce far-field rainfall where and when it was most desirable! In other words, some loss of dryland territory for the gain of a new supply source for freshwater. In addition, marine microalgae may provide a new sustainable food and fuel source (Greene et al, 2016) as well as new aquaculture and commercial fisheries. ***Bridge of Hope Skyscraper,*** winner of Honorable Mention in the 2012 eVolo Skyscraper Competition, presented by Mohammed Adibe and five colleagues is, in effect, a discretized city of barges supporting homes and moisture harvesters, connected by pontoon bridges to span the Dead Sea. **Figure 17**, below (Google Image).

Whilst Team Geographos does not have an unlimited period of time to develop all-inclusive estimates, or accurate forecasts for every factor that impacts establishment costs, or a surfeit of data and information about Dead Sea Depression geophysical conditions, such a reality of unconstrained design may be closer than most Macro-Imagineering professionals perceive because of science and technology's trending progressivity! Such inclusive macro-projects as ***No Man's Land*** (2007) and the quite similar ***Bridge of Hope Skyscraper*** (2012) must be cyborg landscapes in order to actually be effective geophysical-consistent architectural macro-project conceptions! An enlarged Dead Sea simply makes both of these technically-advanced mega-projects BIGGER in geographical socio-economical scope, morphing them from modest installations into truly Rem Koolhaas-style cyborg hyper-architectures!

Gulf of Suez- Suez Canal Anthropogenic Transformations

As of AD 2017, no archaeological expedition had located an ancient cutting that might correspond to a wide canal, the so-called ***Canal of the Pharaohs*** that once connected the Nile River to the Red Sea (Redmount, 1985; Quirke, 2009). Our 21st Century Suez Canal, which extends from 29⁰ 55' North at Suez on the Red Sea to 31⁰ 15' North latitude at Port Said on the Mediterranean Sea, carries seawater and biota northward by way of three lakes: Lake Timsah, Great Bitter Lake and Little Bitter Lake. During the past these present-day lakes were merely low-depressions of Egypt's eastern desert. In ancient times, Lake Timsah was replenished via a distributary of the Nile River during seasonal flooding episodes; historians have confirmed that ancient Egyptians had dug a canal linking the town of Suez with the two bitter lakes. Just guessing, it may be that by increasing the seasonal flood flow volume of the Nile River distributary that fed Lake Timsah, the ancient macro-engineers were capable of true hydraulic macro-engineering—that is, using the freshwater's flow to erode, to mine, the remaining loose soil of a shallow previous human earth-mover [slaves, war captives, criminals, low-wage laborers]-prepared channel linking Lake Timsah to the bitter lakes, and thus efficaciously create a ***Canal of the Pharaohs*** to the Red Sea by a "goaded" Nile River distributary's flood-season freshwater overflow! [The general geographical resemblance to the Salton Sea's unplanned and undirected flash-flood creation by uncontrolled hydraulic flushing action during the early-20th Century in the State of California, U.S.A. is geographically and historically remarkable!]

The Suez Canal's 21st Century owners realize with considered certitude that they must face, not shirk from, a very competitive economic future since the navigation distance via a Northern Sea Route passing through the Arctic Polar Zone from a northwest European port to China-Japan is approximately 40% shorter compared to the route via the old or the New Suez Canal. Responding to this provocation, the New Suez Canal's operators have deepened, widened and in some places excavated a parallel route to expedite two-way ship passages, maximizing their waterway's ship traffic capacity and monetary toll income. The New Suez Canal, because of its dimensions and increased seawater flow, has greatly diluted the bitter lakes, making them ineffective at stopping or screening the Mediterranean Sea from Red Sea biotic invasion (Galil et al., 2017). It may soon be absolutely necessary to interject facilities to alternately bathe passing ships in hyper-saline and hypo-saline waters as an effective means to control the transfer of species from the Red Sea into the Mediterranean Sea, perhaps using quick-cycling floating dry-docks or redesigned, geographically better isolated or multiplied bitter lakes (Brock et al., 1999). Any rise in elevation of the world-ocean as, say, expressed as well as prevalent in the Red Sea and the Mediterranean Sea, of only one-meter would increase the volume of the New Suez Canal's seawater content and, thus, broadened by about 12%, result by the permanent submergence of adjacent landscape. Possible disruptions of the New Suez Canal's daily operation because of landslips followed by necessary clearing dredging, as well as flooding damage to bridges and tunnels passing mobile traffic over and under the New Suez Canal (Soliman et al., 2001).

And, the expected global sea-level rise will only make the impact of tsunamis on both entrances/exists of the New Suez Canal all the more potentially devastating. Whilst reading a fine report on the effect of "worst case" tsunami scenarios for the Mediterranean Sea coastline of Egypt, Team Geographos became aware that it author never mentioned the Old Suez Canal (Hamouda, 2010). Either the author (I) did not think there were any important possible effects on operations and shipping were worth speculating about or (II) the author consciously chose not to annoy anyone in Egypt's government with a speculation supported by geoscience facts of the impressively costly threat clearly posed to Egypt's major income earner, second only to tourism! "The mean recurrence of strong tsunamis in the several basins...[of the Mediterranean Sea Basin]...varies greatly but the highest event frequency (1/96 years) is observed in the east Mediterranean" (Papadopoulos et al., 2014, page 81). By 2012 AD that author's incomprehensible conceptual omission was first corrected by Team Geographos (Finkl et al., 2012) and, during 2016 AD, by another group based in Egypt (El-Eraki et al., 2016). The Egyptian group cited neither Hamouda's 2010 nor our 2012 publication!

Yes, it is an irrefutable fact of R&D life that insufficient, or hasty and cursory, geoscience literature research, by even recognized expert investigators, often leads to a resulting conclusion where, almost literally, the wheel's re-invention is rematerialized....again, and again! Probably the first modern scientific field observations of a tidal influence in a major river, the Columbia River in the State of Washington, U.S.A., was made by B.W. Wilson and A. Torum during their observations of the local fluid aftereffect of the 1964 State of Alaska earthquake. World and regional ocean tidal stages are more important than the volume of river freshwater outflow in determining the impact of a major saltwater tsunami (Tolkova, 2016, page 389). However, the seawater density flow-through at the Suez Canal is virtually ignorable as it is so miniscule in both volume and velocity. It is known that canals and dune-fields paralleling a coastline can mitigate the effects of tsunami wave shore run-up, so it behooves the New Suez Canal's operators to contemplate such mini-megaprojects, perhaps along with movable or inflatable anti-tsunami dams (Shim and Yoon, 2017) situated at both ends within the New Suez Canal, as much needed future protections for their ecosystem-nations' essential source of international financial reward as well as profound geopolitical presence (Abril and Perianez, 2017).

CHAPTER THREE

SECTION II: INTERIOR CORE CYBORG ZONE

Connected through the Strait of Gibraltar [13 kilometers wide, sill depth 290 meters] with the North Atlantic Ocean, the Mediterranean Sea is a semi-enclosed basin, a basin representing only 0.82% of the surface area and 0.32% of the volume of our world-ocean. It also connects with the Black Sea through the Bosphorus [3 kilometers wide, 35 meter sill depth] and Dardanelles straits. The Sicily Strait, with a seafloor sill depth that is a little less than the Strait of Gibraltar's, divides the seawater of the Mediterranean Sea in two sub-basins: an eastern basin influenced by the North Atlantic inflow, whilst the eastern basin is stirred by wind and wind-induced seawater mass formation. (Team Geographos assumes the Suez Canal will become during the 21st Century regulated by sea-locks like Panama's canal.) Oceanographers tend to explain the Mediterranean Sea as a small and representative cul-de-sac "outdoor laboratory", that is, as a spatially-reduced, almost symbolic, mini-version of the world-ocean wherein event-processes can be characterized with smaller, mentally and super-computationally more easily apprehended investigatory geoscientific scales (Goffredo and Dubinsky, 2014). The Dardanelles has a maximum width of 7 kilometers and a sea-bottom sill depth at 70 meters. Thus restricted, only about 300 cubic kilometers of Black Sea water flows into the Mediterranean Sea annually—about two orders of magnitude smaller than that from the North Atlantic Ocean, which is 35,000 cubic kilometers yearly. About 600 cubic kilometers of seawater flows into the Black Sea yearly via the Bosphorus Strait from an intermediary Mediterranean Sea-origin, this intruding seawater is ultimately sourced from the North Atlantic Ocean, contributes salt, heat, and volume-balances the Black Sea. It also ventilates a seawater column at depth and, thus, restrains the upward flux of toxic hydrogen sulfide gas from the Black Sea's bottommost seawater stratum. Seasonal sea-level fluctuations of the

Mediterranean Sea, caused by severe cyclonic storm air-pressure variations (Book et al., 2014), evaporation variations in the western and eastern basins, and North Atlantic Ocean water-mass fluxes, are the dominant forcing mechanism affecting the Black Sea as seawater-channel friction constrains and delays the exchange of seawater through the Bosphorus Strait (Sannino et al., 2017). According to some nudged computer models sea-level will rise by 2100 AD at Ismir and Istanbul by, respectively, 61 and 70 centimeters (Abadie et al., 2016) Now it is necessary for this Macro-Imagineering expose to generally outline prospective mega-project possibilities related to the Black Sea's future anthropogenic infrastructure, taking into account the macro-problem of its enormous hydrogen sulfide gas content and the threat it poses to humans.

Homo sapiens' Cyborg Black Sea Basin: Possible 21st Century Features

"Geomythology" was first neologized circa 1973 by Dorothy B. Vitaliano (1916-2008). Geographical and geological myths have always sparked world-public interest in very ancient, but seemingly universal stories, such as the Great Biblical and Koranic Flood. At circa 7,400 BC, the Black Sea was a freshwater lake whose "sea-level"—the actual volume of water in that particular Earth-crust basin—was then about 100 meters below the sill which separated the post-Ice Age rising Mediterranean Sea from the inland lake. Seawater exiting the Mediterranean Sea flowed through a dryland terrain into the rapidly enlarging future Black Sea as a world-ocean sourced saltwater river which no human could swim (Piper, 2016; Wegwerth et al., 2016; Goldberg et al., 2016); the Black Sea, a mixed-water lake filling to the level of the Mediterranean Sea, may have had a duration of 1,600 years. Because of the difference in freshwater and seawater density, the denser seawater influx naturally pooled at the bottom of the freshwater lake and, subsequently, that underlying body of salty liquid became anoxic, allowing the production of hydrogen sulfide gas below the thin oxic layer. Certain geo-mythologists—Drs. Andrei P. Chepalyga in Russia and W.B.F. Ryan in the U.S.A.—have charmingly supposed that this evidentially-proven factual flooding event at 7,400 BC was the earliest source for the Noah Flood Ur-myth, but others have, with equally geometrical logic, denied that connective theme's geophysical credibility (Roveri et al., 2014; Bikoulis, 2015; Giosan et al., 2012). "Perhaps the most geologically intriguing [geomyth] is the Greek story of Dardanus, who is said to have founded Troy after surviving a flood that occurred when a barrier in the Bosphorus/Dardanelles gave way, allowing water from the Black Sea to rush into the Mediterranean and temporarily inundate the island of Samothrace.... Except for the fact that water is flowing in the opposite direction, this story bears a striking resemblance to the [process-] event that flooded the Black Sea around...[7,400 BC]" (Tepper, 1999, page 223). [Actually, the story fits a tsunami scenario.] The

Black Sea's rapid filling, to its present-day level, aggravated our planet's Chandler Wobble—the free motion of the pole of rotation around the main inertia axis of Earth discovered in 1891 by Seth Chandler (1846-1913); that process-event could be responsible for the most drastic pre-Space Age ceno-Anthropocene change in the rotational parameters of the Earth (Spada et al., 1999).

Herman Sorgel's grandiose macro-project, ATLANTROPA, reducing the Mediterranean Sea's seawater level, by 200 meters or more, through natural evaporation of an isolated basin might, if ever undertaken as initially planned may induce a similar effect owing to the rapid redistribution of Earth's seawater mass over two centuries (Anzidei et al., 2014). For hydraulic loads with a horizontal geographical extent of less than 2,500 kilometers, a strong localized signal can be measured as the inhomogeneous Earth-crust's elastic response becomes detectable. There are about 50 sub-aerial and submarine volcanos active or quiescent in the Mediterranean Sea Basin. Named for a geologist, Luigi Ferdinando Marsili (1658-1730), the Marsili Seamount aptly represents the man for whom Europe's biggest volcano, larger than Mount Etna, a geoscientist who pioneered geological mapping (Romano et al., 2016). The seamount is an active volcano, discovered by oceanographers during the 1920s, which must be continuously monitored and the hazards related to explosive ongoing and future eruptions properly evaluated (Tamburrino et al., 2016). [This cautionary approach is required because of past and ongoing volcanic eruptions at the infamous Greek island-caldera of Santorini (Nomikou et al., 2016)!] Some macro-imagineers foresee a time when the Marsili Seamount becomes a useful geothermal energy resource for the region's geographically expanded and intensified human infrastructure (Italiano et al., 2014).

On 9 September 2014, in Norway at the coastal site of Molde, emergency response workers investigated an unpleasantly foul odor in the ambient air, but without conclusion as to its source. That rotten eggs odor spread eastwards, later reported in Sweden and as far distant from coastal Norway as Muonio, Finland! The cause for the repulsive odor was later determined to be hydrogen sulfide [SO_2], probably originating from an eruption of Iceland's Bardarbunga volcano (Grahn et al., 2015). Smart oil-field and natural gas infrastructures as well as mining complexes are being multiplied in the Black Sea-Caspian Sea region (Zhiltsov et al., 2016). Fast long-distance road and railway new connections with Russia and China are being planned by macro-engineers. At Istanbul, Turkey, three suspension bridges currently span the Bosphorus (Geymen, 2013) and a fourth suspension bridge is planned at the Gallipoli-Lapseki worksite to span the Dardanelles by 2023 AD. The completed and operational Marmaray subway mega-project serving Istanbul has been tunneled beneath the Marmara Sea's busy international waterway (Koprulu et al., 2016). Even an anthropogenic canal, the so-called Kanal Istanbul Macro-project, that offers the attractive prospect of a second international shipping channel between the Black Sea and the Mediterranean Sea has been seriously advertised (Orhon, 2014). Along with the eastern Mediterranean Sea, the Black Sea's basin is shaken by powerful earthquakes [Azov Sea 1650 AD to Izmit Bay 1999 AD] that generate tsunamis (Cemen and Yilmaz, 2017). More such temblors must be expected during the 21st Century—it is simply a fact of life for civilization in that region of the Earth's crust. Undersea pipeline construction, submarines moving at high speed, large sinking vessels, small asteroid impacts, large-scale underwater explosions as well as Nature's earthquakes can disturb the lowermost toxic hydrogen sulfide-saturated seawater layer which rests atop the Black Sea's toxic marine sediments and shale rock. Massive seawater stratum disturbance does pose a real continuing existential threat for humans, and other living creatures, situated anywhere near the shoreline of the Black Sea in the form of a short-term poison-gas cloud envelopment following, perhaps, tsunamis. If the gas can be smelled very strongly, then precautionary health-preservation reactions must occur immediately since a strong concentration of the inhaled gas is fatal!

The Tunguska airburst of 1908 AD and the AD 2013 Chelyabinsk 500 kilotonne TNT equivalent mid-air bolide burst event furnished humanity with ample warning notice of what is, inevitably, to come to our homeworld in near-term future Earthly Geologic Time (Rumpf et al., 2017). On 6 June 2002, north of Libya and between Italy and Greece, an asteroid burst apart high above the Mediterranean Sea, with an explosive yield of about 26 kilotonnes of TNT. Air bursts deposit all energy into the atmosphere since they do not reach the land or ocean's surface. Twice as many asteroids, as airbursts mostly, strike near the ocean's surface as compared to the number exploding above or actually hitting our planet's landmass. Over seawater airburst asteroids, because of physics, do not pose significant tsunami water-wave effects! Those airbursts happening over seawater endanger few persons—only those exposed aboard surface vessels or surfaced submarines—with short-range airburst thermal and wind effects. Tsunamis generated by asteroid impacts on seawater are less dangerous to persons than might be commonly intuited since the saltwater wave run-up past the beach shoreline, onto the recreational or industrial landscape, usually is only a relatively short distance (Rumpf et al., 2017). For residents and workers near the coast, the mortal threat of poisonous gas eruption from the Black Sea caused by asteroid impact is the most ominous and frightening (Badescu et al., 2010). The energy released by the Tunguska event has been estimated as equivalent to the release by ignition of ten million mega-tonnes of TNT explosive; an impact of that magnitude in the Black Sea could create a hovering steam cloud bearing poison that might cover 105-210 square kilometers, depending on ambient wind strength and direction, affecting about 26,000 to 52,000 persons onshore (Badescu, 2009). [Note: "detonation" is only the sound that issues from any explosion, the "boom" or "bang".] The impact's follow-on tsunami may be determined for any shoreline location by using simple ray-tracing to obtain an approximation of the affected coastline, depending on Black Sea bathymetry and the landscape's natural and anthropogenic topography. With the instantaneous release of energy equivalent to 10 mega-tonnes of TNT, a tsunami wave-height of 58-62 meters locally is expected, diminishing in height until the main wave approaches the nearby strand by passing over the continental shelf. Who might explode a nuclear weapon on, or in, the Black Sea? Reportedly, Russia has designed, and is building, drone submarines in its **Ocean Multipurpose System "Status-6"** weapons-delivery system program. Or, Iran might seek to use a powerful bomb, whatever the type and nature of its explosive power, within the Black Sea for any number of nefarious reasons. Still, it seems to be Nature that really is the most obvious threat to lives of people settled on, or frequenting, the Black Sea's shore. If the stratification of the seawater column weakens by further thinning of the sea's oxic layer, then biota within and around the Black Sea becomes increasingly endangered (Capet et al., 2016; Churilova et al., 2017), possibly even to regional extinction (Kump et al., 2005). Holiday crowds during the summer [June 26th] and autumn [12th September] of 1927 were startled and alarmed by the sudden eruption of biogenic methane/hydrogen sulfide gas-fueled flames tens of meters high over the Black Sea offshore Sevastopol, Cape Lucullus and Yalta occurring simultaneously with a series of strong earthquakes focused near the Sea of Azov. So, there are recorded precedents for suspicion that toxic hydrogen sulfide, in the form of a traveling cloud, is a real-world concern! Team Geographos first became aware of this still unique macro-problem for those persons dwelling and working in the vicinity of the Black Sea by reading A. Spiridonov's "Saving the Black Sea", ***Science in the USSR***, Number 6, November-December 1990, at pages 51-58. As yet, Macro-Imagineering has only a few Nature-corrective mitigation technological "arrows" in its quiver of ideas. Cesare Marchetti (born 1927), working at the International Institute for Applied Systems Analysis in Laxenburg, Austria has suggested a bioengineering approach to drawing down aerial carbon dioxide gas—that is, growing some macroscopic green plant on the Black Sea's surface whose remains can sink to the sea-bottom to be interred by the anoxic waters but that does not solve the macro-problem we are addressing here. Our world's largest reservoir of toxic hydrogen sulfide is the deepest seawater layer of the Black Sea! Hydrogen sulfide potentially has economic value if both sulfur and hydrogen can be recovered cheaply. Any concept of extraction of dissolved hydrogen sulfide from seawater is indisputably a radical Macro-Engineering means of diminishing the major biota-threatening environmental macro-problem of the dormant, potentially lethal Black Sea. Nevertheless, questing macro-engineers have formulated some interesting mega-project plans, many of which foresee a worldwide Hydrogen Economy impending: (I) Midilli et al. (2007, page 117) have conservatively calculated that hydrogen capture might produce "...38.3 million TJ [Terajoules] of thermal energy or 8.97 million GWh [Gigawatt/hours] of electricity energy"; (II) Naman et al. (2008) proposed a pilot-plant for removing hydrogen sulfide from seawater; (III) Haklidir et al. (2009) reiterated that, if economical chemical decomposition processes are achieved, then the Black Sea will become a valuable resource for our world's future Hydrogen Economy; (IV) Azarenkov et al. (2014) proposed a massive floating electricity generation facility and (V) Beschkov et al. (2015) offered a sulfide-driven fuel-cell to do the required harvesting chemistry! Industrialization of a semi-enclosed part of the world-ocean at such an expansive geographical scale is bound, non-mysteriously, to make that increasingly wealthy region a "self-powered" cyborg simply by utilizing the Black Sea as a kind of Nature-made battery!

Crust-depression Power Generation Macro-projects: Beneficiaries of Sea-level Rise

Djibouti's Lake Asal aside, Africa's biggest natural crust negative indentation is the teardrop-shaped Qattara Depression of Egypt. Although it has six semi-permanent saline lakes, the best known is Moghra Lake, which has an area of about four square kilometers (Khan et al., 2014). The Qattara Depression's true geological formation is still hotly disputed but it suffices to say only that it is a capacious below-sea level Nature-made pit with an area but 2% of Egypt's total territory which, if filled with seawater brought in from our AD 2017 Mediterranean Sea, could contain 1,340 billion cubic meters of Nature's pervasive fluid (Hereher, 2015), about ten times the live storage of the Aswan High Dam Reservoir's capacity. Aerial evaporation in the desert climate regime would eventually fill the Qattara Depression with dense salt, after fifty-eight seawater "equal doses" had been liberated into the vast aridic pit—that is, after cycling 77,720 billion cubic meters from the world-ocean to the Qattara Depression. This would take a few years to transpire! **Figure 1**, below, Google Image illustration of the Qattara Depression filled to AD 2017 sea-level.

The geomorphology between the 20,800 square kilometer Qattara Depression and the Mediterranean Sea, famous for the fierce World War II turning-point land-sea-air battle at El Alamein in October 1942 (Bondesan et al., 2013), is suitable for seawater conveyances such as canals, tunnels and pipelines (Hafiez, 2011), although many World War II minefields are still unremoved! Western Egypt, approximately 248,000 hectares located 60 kilometers west of Alexandria, are seeded with millions of items of unexploded and still life-threatening World War II ordinance. [A propagandistic Hollywood film, ***Sahara*** (1943), starring virile American actor Humphrey Bogart (1899-1957) somewhat memorialized this offensive armed conflict victory of the once-besieged Allied armies in northern Africa.]

Future postulated global mean sea-level rise should make this power-generation mega-project even more feasible economically than it already is assessed to be by hydropower development experts! Team Geographos' member Richard B. Cathcart prefers merely to fill the indentation to current sea-level, whatever that may be by mid-21st Century, and to float large barges, presenting photovoltaic apparatus to the Sun's rays, sending electricity to Cairo, supplementing the power output of the Aswan High Dam on the Nile River. Such barges would not be subject to ship collisions, big storm-generated sea-waves with long fetch and could be guarded more easily than land-based facilities for state security reasons; seawater coverage by deployed barges should effectively reduce the evaporative surface area, slowing the rate of sea-salt buildup in the basin! Resiliency of civilization's infrastructure! Why not name this anthropogenic seawater feature "New Lake Moghra"? [The International Mass Loading Service, established AD 2002 by the NASA to monitor Earth-crust deformation caused by natural and unnatural redistributions of air and water, will surely track the ongoing effects of hydrological loading at a fast-forming "Lake Moghra" as well as the planned Dead Sea enlargement!] The localized meteorological effect of the lake's existence, and its annual 2.7 meters evaporation, should be primarily seen as existing dew-fog enhancement, with little effect downwind on the lower Nile River Valley where Cairo sits. And, of course, all seawater extracted and delivered to the Qattara Depression will be compensated by greater inflow through the Strait of Gibraltar from the North Atlantic Ocean. "Egypt now releases less than 10% of its water supply, a mostly saline and highly polluted aqueous mix, to the sea, with little sediment available for coastal replenishment" (Stanley, 2017, page 5). Because of the Aswan High Dam and other dams farther upstream, so little freshwater enters the Mediterranean Sea via the Nile River that it might be worthwhile for Egyptian macro-imagineers to consider placing main railway and highway routes, nowadays paralleling the Nile River but obscuring valuable farmland, into the middle of the Nile River. Dalia M. Gouda (2016) opts to relocate Egypt's industries to make use of the desert, freeing the Nile River Valley landscape for agriculture.

In Hubei Province, central China, the economically thriving town of Gufuzhen in Xingshan County has been connected since 2015 AD to the super-highway linking Shanghai and Chengdu in southwestern China via a 10.5 kilometer-long roadway on a bridge built generally following the water flow's centerline of the Xiangxi River! Incidentally, architect Paul Maymont (1926-2007) during 1962 proposed a utilities-corridor, subway and highway tunnel exactly following the route of the sinuous scenic Seine River (Chen et al., 2014) flowing to the North Atlantic Ocean through beautiful downtown Paris, France. **Figure 2**, below, a Google Image, illustrates Maymont's wonderful unrealized architectural concept.

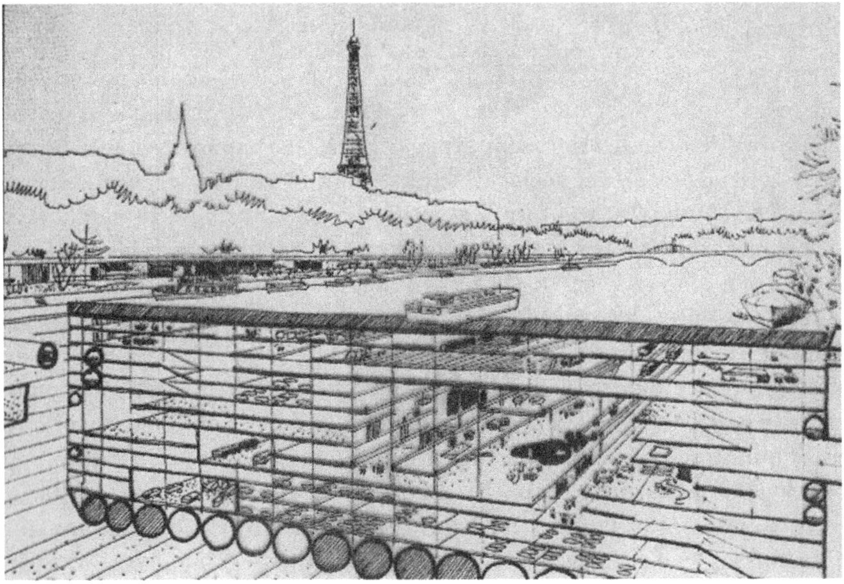

"Treeing the CATS: Artificial Gulf Formation by the Chotts Algeria-Tunisia Scheme", penned by Nicola M. Pugno, Richard B. Cathcart and Joseph J. Friedlander, pages 489-517 in Viorel Badescu and Richard B. Cathcart's editorial landmark ***Macro-engineering Seawater in Unique Environments: Arid Lowlands and Water Bodies Rehabilitation*** (2011) described a mega-project meant to redeem the almost rainless wastelands of western Tunisia and southern Algeria using big seawater inundations permanently linked to the Mediterranean Sea by an east-west shipping canal facing the tidal Gulf of Gabes (Othmani et al., 2017). Two masterful Swiss stratigraphers, Pierre Jean Edouard Desor (1811-1882) and Hans Conrad Escher von der Linth (1807-1872) had proposed in 1863 that a diluvial "Sahara Sea" was the cause of the Northern Hemisphere's newly-discovered Ice Age period of past Earthly Geologic Time! Quite aside from the remote possibility that their theory has any validity whatsoever, Jules Verne (1828-1905) penned a story, ***Invasion of the Sea***, that aroused extra-ordinary world-public ardor for inundating mostly deserted low-lands, especially those then known to exist in the awe-inspiring Sahara, a vastness of aridity just a short distance across the Mediterranean Sea from lushly green southern Europe—in fact, Verne's expeditionary novel was explicitly mentioned by Rosalind Williams' ***The Triumph of Human Empire*** (2014), her popular history of Homo sapiens' dominion over Earth's Nature. 20th Century macro-engineers such as J.B.F. Champlin, J.W. Poston and J.A. Lake offered their "Project Pecos Bill" in their 14-16 January 1970 report, ***An Interior Seaway for Northern Africa***, at the Symposium on Engineering with Nuclear Explosives held in Las Vegas, State of Nevada, later published, at pages 1643-1660, in ***CONF-700101 (Volume 2)***. During the time of the ancient Greeks there was a natural lake situated in the region just west of the present-day City of Gabes, Lake of Pallas. By AD 43, when the Roman geographer Pomponius Mela (presumed to have died circa 45 AD) wrote a description of Lake Pallas for his ***De situ orbis libri III***, it had dried up and become all but forgotten even by the local populace of paupers! However, the physiography of northern coastal Africa's chotts region remains today what it was then, an available career-making opportunity for modern Macro-Imagineering's professionals! [Jules Verne's tale, by the way, was inspired by a bizarrely advocated macro-project idea for the creation of "la mer interieure" through filling the largest chotts with seawater flowing in short excavated canals without sea-locks, that had previously emanated from Francois Elie Roudaire ((1836-1885) (Roudaire, 1874; Pearce, 2002; Wu et al., 2017). **Figure 3**, below.]

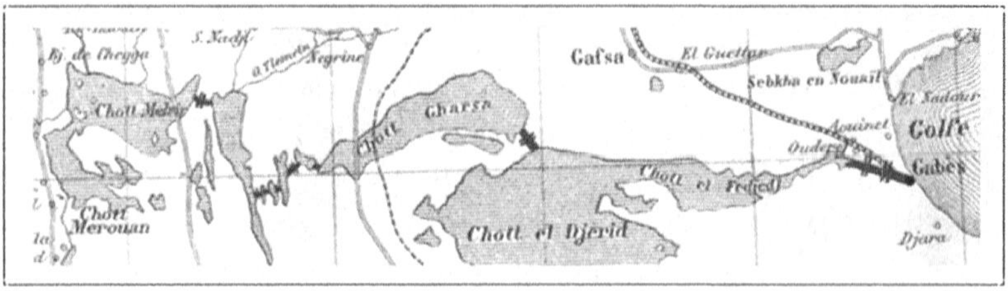

The Orphic Story of DESERTEC: Aborted Mega-project Dream

Our ceno-Anthropocene should be humanity's Era of Infrastructural Dreaming, an open-ended historical period when informed and inventive humans, worldly humans, attempt an epochal global infrastructural upgrade! No ecosystem-nation yet owns a strategic electricity reserve—that is, a very large-scale means of storing electricity—and there is no back-up for the network of long-distance cables called the "international macro-grid". These assertions have a memorable actualized proof since 2018 AD marks the 15th anniversary of the 28 September 2003 blackout—the total loss of electrical power—of an entire 21st Century high-technology ecosystem-state, Italy, caused by two green trees in Switzerland falling onto long-distance electricity transmitting wires linking multiple generation and consumption systems (Byrd and Matthewman, 2014)! A

French engineer, Georges Viel, was the first person to present a macro-plan, during 1929-1930, for a high-voltage electricity transmission macro-grid (400,000 volts) to serve the entirety of then fitfully modernizing Europe (Legendijk, 2008, page 80). Currently, Europe imports 96% of its uranium, 85% of its oil, 67% of its natural gas and 40% of its coal. DESERTEC was founded in 2003 AD to build a renewable power system, derived exclusively from northern Africa solar-energy farms, conveyed to southern Europe via High Voltage Direct Current (HVDC) connections, through its DESERTEC Industrial Initiative of AD 2009; however, for many reasons, some of a distinct geopolitical nature, by 2012-2013 AD the DESERTEC group had been reduced to a minimalistic consultancy status. Instead, another group of Sciencеers, GEIDCO [Global Energy Interconnection Development and Co-operation Organization], has emerged into the foreground of macro-planning. GEIDCO is People's Republic of China-based, has agreements of intent with energy-producers in South Korea, Russia and Japan as well as with public-sector and private-sector utilities, equipment manufacturers and universities in more than fourteen countries. GEIDCO's goal is to have intra-continental macro-grids operational on each continent by 2030 AD and to have inter-continental links operational by AD 2050. Such a global macro-grid would ensure that all, or nearly all, of the electricity generated will be consumed after transmission through +500 kilometer-long HVDC transmission systems because HVDC makes possible the transfer of electricity between macro-grids that are operating at different frequencies.

MEDGRID, although remarkably similar to DESERTEC, and possibly coordinated by GEIDCO, may develop a solar-power generational capacity of 20 Gigawatts, of which 25% would be exported to southern Europe through submarine cables (Worzyk, 2009). [GOTO: http://www.medgrid-psm.com/en/.] On 4 February 2016 Phase 1 of the world's largest concentrated solar power (Muller-Steinhagen, 2013) facility, using one-half million 12 meter-tall parabolic mirrors, was switched "on", activated; NOOR-1, located close to the Sahara in Morocco's impoverished Ouarzazate Province, has a generating capacity of 160 Megawatts. It will be joined by NOOR-2 and NOOR-3 in 2018 AD if all goes according to well-laid macro-plans. So, sometime circa AD 2020, the completed facility will be capable of generating 580 Megawatts, or about 40% of Morocco's energy needs will become resourced from renewable energies production. Yet, no electricity is expected to be exported to Europe, although perhaps shared is the more correct term. There are two 25 kilometer-long AC transmission cables, a cable of 400 kilovolts that was the first electricity cable link between Europe and Africa (Anon., 1999) and another power-cable carrying 1,400 kilovolts that has been operating since AD 2006. Presently, the two interconnections are used to import about 4.2 Gigawatts/year of electricity mainly made in Spain. These cables exit Spain at Puerto de la Cruz in the vicinity of the town of Tarifa and enter Morocco at Mellousa in its Tetouan Region. Northern Hemisphere solar-power generation can only become more efficient as Earth's dryland latitudes migrate northward because of the still inexplicable geographical expansion (by spatial widening) of the Earth's Tropic Zone (Heffernan, 2016). From 31 August through early September 2015, a huge dusty sandstorm blanketed large swaths of seven ecosystem-countries (from the area between Syria and Iraq to the Persian Gulf nations, thence back to Syria, Lebanon and out over the Mediterranean Sea (Parolari et al., 2016). The popular Green media reporting spread fear by introducing the beguiling, but scientifically erroneous geographical concept of a "New Dust Bowl" commencing in that hot dryland region. It is a fact that heat-waves are far more likely to adversely affect people of the region owing to ongoing climate change (Lelieveld et al., 2016).

Team Geographos' Really, Really BIG Sahara Tent!

"By focusing on artistic composition, circuses have reinvested in the relationship with their audience by treating them not just as bystanders, but also as Art lovers" (Loring, 2007, page 17). Team Geographos was first to hypothesize, during 2004 AD, a tentative Scienceer tenting of the Sahara, in some sun-shaded covered places air-conditioned by solar-powered mechanical cooler installations (Cathcart and Badescu, 2004)! Saharan silver ants (Cataglyphis bombycina) forage in the intense mid-day sunshine, when ambient temperatures can be 70^{0} Celsius, even though their body temperature cannot exceed a critical thermal maximum of 54^{0} Celsius; silver ants accomplish this life-preservation feat by combining exterior optical reflection and radiative heat dissipation (Shi et al., 2015). Of course, such a schema might be applied to mobile exploration or exploitation vehicles roaming the Sahara, but also Team Geographos' enormous enclosed volume tent—that is, its landscape-covering reflective fabric material. The solitary acacia **Tree of Tenere** growing in northern Niger's sand-dune landscape grew in the Sahara for more than 300 years until fatally molested by a criminally culpable motor-vehicle driver (Connelly, 2012-2013)—thus, what to us is a shades-of-brown wasteland today, once may have been otherwise vegetated by verdant trees, shrubs and grasses! At an altitude of 2,000 meters in Algeria's Tassili Plateau, the rare Sahara cypress (Cupressus dupreziana) first described botanically in 1924 AD, survive in a grove of about 200-250 known tree specimens (Werner, 2007).

Green, the color, is a term not previously misused as it has been since circa AD 1979 by radicalized persons! Macro-Imagineering offered during 2004 the macro-project equivalent of a non-gallery work of Art, namely Agnes Denes' ***Tree Mountain—A Living Time Capsule*** (1996 to the present) (Chianese, 2013; Xue and Shukla, 1996). The logical lapses and empirical deficiencies of the popularized, 1920s-1930s-originated, geographical concept of "desertification" have been known for decades but, nevertheless, this mistaken and misleading notion has been radical Green media-institutionalized globally since circa 1970 AD (Behnke and Mortimore, 2016). With regard to the Sahara, part of the macro-problem is that archaeological evidence has not been included in debates about flora and fauna conservation priorities (Foley and Lahar, 2015; Boivin et al., 2016; Wright, 2017). Other than recession of glaciers since the most recent Ice Age, the most prominent environmental change of the past 6,000 years is the increasingly better understood transition of northern Africa from a well-vegetated "Humid Period" (Boos and Korty, 2016) dominated by grasses in association with scattered acacia trees into our Earth-bioshell's largest hot desert. However, the vast Sahara dryland has existed for seven to eleven million years, formed as the ancient Tethys Sea shrank, especially at the Gibraltar Arc region (Crespo-Blanc et al., 2016), and warm desert conditions subsequently expanded geographically across northern part of the African continent (Zhang et al., 2014). Gibraltar Arc is an exceptionally narrow-gapped landform. **Figure 4** (Google Image).

Figure 3. From a viewpoint looking eastward into the Mediterranean Sea Basin, a satellite image snapped from a high-elevation oblique vantage point above the eastern North Atlantic Ocean. The clarity of this Space Age image makes it so popular that it was used, amongst other publications, as an attractive, colorful front-cover book sales-promotion picture by Pierre Beckouche (Ed.) in ***Europe's Mediterranean Neighborhood: An Integrated Geography*** (2017, Northampton, Massachusetts: Edward Elgar Publishing. 277 pages).

Published scientific speculations, few of which are internally consistent conceptions, concerning our Earth-bioshell's far-future geophysical status are legion in number already. The amount of energy being radiated by our Sun seems to be increasing approximately 1% per 110 million years. One hundred million years from now, it has been supposed, another supercontinent will aggregate around the Arctic Zone, forming "Amasia" where a re-shaped, mostly elongated, Mediterranean Sea might still exist amidst the world's conglomerated tectonic plates (Mitchell et al., 2012)! [Afterwards, supposedly, all liquid water will disappear from Earth's face in about one billion years (Leconte et al., 2013). In other words, Earth will be all landscape all the time, and rather like Willem Goeree's unique 1694 AD cartographic appreciation of an ocean-less Earth presented in **Figure 4**, Section I of Chapter Three.] In today's almost maniacal ceno-Anthropocene eco-socialist era of politically-correct prognostications, predictions are wanly issued through the popular "news" media which seem always to predict only future disasters awaiting humanity; some geoscientists, seeing a market hungry for such exciting forecasts, intending to garner for themselves some world-public recognition to assuage their egos and possibly fatten their personal bank-accounts, no longer are reticent about furnishing the geoscience ignoramuses populating 21st Century "news" media with suitable fodder; sadly, the worst part of this eco-sociological situation is that professional "news"-persons readily assume an intolerant mask of superiority over all their readers and viewers, mindlessly propagandizing fresh disasters unendingly! Will the rate of spreading at the Mid-Atlantic Rift separating tectonic plates increase drastically (Marques and Kaus, 2016)? Do powerful recorded earthquakes which devastated Portugal in 1755 AD and 1969 AD indicate a new phase in the Gibraltar Arc's behavior has begun (Duarte et al., 2013)? It is known, with a fair amount of certainty, that Western Europe, including the British Isles, as well as the eastern Mediterranean Sea Basin are, currently, subsiding Earth-crust segments, whilst the western Mediterranean Sea Basin is not currently known to be subsiding (Ostanciaux et al., 2012).

MEDICATIVE MACRO-IMAGINEERING: Earth + Mars Megaprojects (2014), authored solely by Richard Brook Cathcart, in Chapter 8, "Mediterranean Sea Basin Enclosure and Development", offered a compleat modernization and explication of a feasible 21st Century cyborg ATLANTROPA macro-project plan empowering the region's people surrounding a stabilized cyborg Mediterranean Sea able to cope with a future one-meter higher mean global sea-level, then possibly prevalent everywhere else by 2050-2100 AD. However, in the few years since, some new geologic and historic facts emerged that warrant elucidation, therefore emboldened and invigorated Team Geographos considers it quite necessary, and indubitably appropriate, to continue the alluring cyborgized ATLANTROPA's important saga, so to speak! Whilst the Sahara's silver ants reflect sunshine and therefore, on a tiny geographical scale, alter the immediate weather pattern and climate regime, human cities generate waste heat emanating from everyday human activities that is known to broaden the Northern Hemisphere's jet-stream and alter other major Earth-atmosphere systems (Zhang et al., 2013). The Sahara covers a large area, approximately 9,400,000 square kilometers; if that almost unimaginably vast area's albedo were anthropogenically changed by, say, a darkening caused by new irrigated green plantation vegetation, or a vast reflective Sahara Tent, then there might be a regional reduction in surface air pressure, a strengthening of the trade winds, the westerlies and the North Polar Zone easterlies, which in turn should increase meridional heat transport to the Arctic's air. Possibly the Sahara may, in future, be subjected to markedly increased snowfall were the jet-stream shifted southwards oftener (Galvin, 2010). [Snow fell on southern Algeria, at Ain Sefra, on 18 February 1979, 19 December 2016 and 20 January 2017.] Paleo-rivers that in the forgotten past flowed gravitationally south to north through the wetter climate regime times of the Sahara were once major Northern Hemisphere human migration routes during northern Africa's "Humid Period" (Coulthard et al., 2013; Nash et al., 2016). During our present-day's dryer Sahara climate regime, humans are mostly reliant on mined groundwater to survive and thrive (Goncalves et al., 2013), of which there is a huge remaining potential supply generously provided by a motherly Nature (Abotalib et al., 2016; Trezzi et al., 2017). At the coast, future development must adhere to sound macro-planning principles in order to avoid any contaminating groundwater flood caused by intruded saltwater associated with global mean sea-level rise.

Dust storms, blowing from the Sahara, particularly during Winter and Autumn, can temporarily alter the dynamics of the Mediterranean Sea's surface seawater layer through large induced increases of chlorophyll production fomented by dust deposition on the seawater. It is anticipated that, in future, Mediterranean "hurricanes"—also known as "Medicanes"—will be more powerfully present over the Mediterranean Sea than during AD 2017 (Flaounas et al., 2016). A sea stirred by air's movement, however, is better oxygenated. "In the northern Sahara, meteorological records show a tenfold increase in dust storms during the period when World War II military campaigns disturbed fragile desert vegetation and soils" (Machlis and Hanson, 2008, page 731). Naturally, if fully tented, such a bio-oceanographic effect could never again occur! In the battles of North Africa against German Field-Marshal Erwin Rommel (1891-1944), his opponent, English General Archibald Percival Wavell (1883-1950), had special maps which showed the probable movement of aeolian sand-dunes during certain seasons. The only major coastal macro-project promoted by Benito Mussolini (1883-1945) in post-1920 conquered and colonized Libya was the Litoranean Highway, which he personally dedicated during 1937 AD, that was supposed to connect Tunisia to Egypt; by the late-1940s the westernmost section was impassable, buried by migrating aeolian sand-dunes (Heefner, 2015)! Ignoring vortex winds, slower non-circular winds that do not lift individual sand grains and dust particles off the ground do move and reshape dust piles and sand-dunes sitting on the ground laterally by saltation. And, for this reason, the blind French Sciencer and radio amateur Paul Remy (1879-1948) insisted that a subway was necessary for the future industrial development of some Saharan places—for example, he proposed that a 202 kilometer-long segment of a 1,610 kilometer-long railway, extending perhaps from near the inland city of Smara, Morocco, through Mauritania's national-territory to fabled Timbuktu [16^0 46' 33" North latitude by 3^0 00' 34" West longitude] located in Mali, was originally intended ultimately to link Nigeria and the mineral-rich Congo River Basin beyond, since it must pass through a large area of unfixable shifting sands, should be done by trains operating underground, in a subterranean two-track tunnel (Anon., 1929, page 34)! Nowadays, if that subway were extant, it might be connected by its owners to westernmost Europe via a Gibraltar Strait Railway Tunnel (Pliego, 2005) across the Gibraltar Arc landform's water-gap. Tunnels (hard-rock, immersed, submerged floating), floating pontoon fixed crossings and a long-distance suspension bridge have been planned, even designed for expedited high-speed passages by freight and passengers across the Gibraltar Strait.

Tenting the Sahara, or at least a large part of it, is a housing infrastructure that is intended by Team Geographos to be used by significantly large numbers of people for practical purposes. Although it can never be an exact match to every human's needs, it can alleviate disappointment and encourage hope by eliminating "homelessness" (Larkin, 2013)! Actual tent "fabrics" might be variable, perhaps dynamic architectures such as foldable short-term domes, safe and stable self-assembling granular materials [aleatory architectures (Keller and Jaeger, 2016)] and novel physicalist urbanisms like those proposed by Magnus Larsson (2014). [A stimulating preview of their possible useable artworks composed of sand and biota [bio-grouts] can be viewed, since AD 2012, at Tottori, Japan's unusually entertaining **Sand Museum**!] Even the dream kinetic artworks-machines devised by Theo Jansen of The Netherlands could cultivate sheltered new agricultural landscapes, formerly aeolian sand dunes that have been sprinkled with nanoparticle fertilizer (Kottegoda et al., 2017) and pure or impure freshwater—both might be systematically applied by overhead spray-dispensers equipped with special nozzles perfected for cloud brightening mega-projects such as increasing the reflectivity of marine stratocumulus clouds (Salter et al., 2014)! Such uniquely designed nozzles were suitable for spraying seawater into the air over the open-ocean, so there should be no difficulties with a mixture of highly comminuted fertilizer particles and fluid freshwater. Ah, but from whence the vital liquids? In AD 2008, Team Geographos suggested dousing the most mobile desert sands of the uncapped Sahara with seawater extracted from the ocean and then pumped long-distances via pipeline into the most arid segments of northern Africa and, later, even proposed tapping the Amazon River's wasted freshwater outflow plume as a sustainable source of vital liquid capable of being transported across the South-North Atlantic Ocean to Mauritania by buoyant submarine pipeline, thence inland to the middle of the Sahara's arid vastness (Badescu et al., 2010).

Mediterranean Sea: Dr. J.E. Dobson's Aquaterra Incognita Preserved

The 13 kilometer-wide Strait of Gibraltar is Nature's only gap in the topographic barriers separating the 2.5 million square kilometer Mediterranean Sea from the world-ocean. Cigar-chomping Herman Sorgel's post-1928 AD, continually revised a mega-project plan to fill that geographical gap to foster his vision of ATLANTROPA. It was not a gag (Lehmann, 2016). [Ed Regis's ***Monsters: The Hindenburg Disaster and the Birth of Pathological Technology*** (2015) overlooked Herman Sorgel's possibly horrifying—well, at least, astounding—ultra-project earliest version of ATLANTROPA, not *CYBORG ATLANTROPA.*] Even under warmer climate regime scenarios, the Mediterranean Sea's well-managed landscapes are less prone to wildfires, so-call "megafires" (Bowman et al., 2017). Next, the abundance of plastic debris contaminating the seawater, which was first reported by R.J. Morris (1980), must be better managed, strained from the water and removed. It is entirely possible to mind-fancy the Mediterranean Sea as a gigantic well-managed anthropogenic oceanarium, a magnitude larger than the aquaria that may be the future Qattara Depression and Dead Sea! On the European side of the strait, the famed and unremittingly disputed ownership of which is an outcome of wars and legal proceedings, the Rock of Gibraltar is the focus of stronger wave energy than any other strand site in the Mediterranean Sea Basin and, consequently, enduring continuous shoreline retreat whilst on the African side of the strait, sits the Rock of Gibraltar's geographical counter-part, El Acho, a part of the Ceuta Promontory. Yearly, approximately 110,000 vessels use the split shipping lanes through the Strait of Gibraltar. During the past decade, the ships that are the network of our world's economic trade have become truly gargantuan, with bulky payloads to match their size. So, since 1898 AD, sea-lanes—charted paths that regulated ships must follow—have been established by concerned international groups in order to forestall major 21st Century ship collisions, losses of human life and ships foundering. In the Strait of Gibraltar, the ship-traffic separation scheme demands navigational skills now mostly wielded by nearly-AI robotic navigators monitoring the Automatic Identification System broadcasts from all nearby ships! With crewless robotic ships soon to be present on the world-ocean, it is now almost a given that steersmen are a dwindling breed of vessel pilot!

Many Sci-Fi and fictional historical dramas center on the theme of human and Earthly landscape cleansing, as well as judgment of landscape inhabitants, by great floods wherein unstoppable water flows temporarily inundate or permanently submerge cherished swaths of landscape. The supposed impending threat of ocean surface elevation has increased the world-public's interest in studies of submerged prehistory and past, Nature's previously unrecognized sea-level fluctuations, as 21st Century organizations seek involvement to preserve economic interests such as costly heritage and modern infrastructures (Paris et al., 2016; Harff et al., 2016). In the real-world, the American geographer Jerome E. Dobson (2014) correctly dubbed those landscapes currently obscured under an impermanent seawater layer as "***Aquaterra Incognita***" and pointed out that those places may reappear sometime in the future just as they disappeared in the past! Or, if we successfully exclude from the Mediterranean Sea Basin any future global mean sea-level rise (Galassi and Spada, 2014), monitored with globally collected data since 1933 by the Permanent Service of the International Council for Science (Rickards et al., 2015), humans will have preserved that unique watery Earth-bioshell regional feature and its most adaptable creatures! Published last year, an affective geographical scenario projection to 2100 AD of the effects of a non-excluded global mean sea-level rise (focused on Lipari Island owned by Italy) offers a carefully-worded report which definitely gives some good basin-wide modelling indications of the expected impact of a future regional marine submergence, especially in terms of the hazard implications for all humans situated near the Mediterranean Sea's present-day shoreline (Anzidei et al., 2017). Thrifty Swiss investors may wish to reconsider their tangential association with Sardinia's virulently political secessionist social movement which has sought to become officially a part of Switzerland since AD 2015 (Albanese and Letzing, 2015) since their "Canton Marittima" property could be spacially diminished in area by significant regional sea-level rise; real-estate speculators never knowingly pay for more landscape but then occupy less landscape because of Nature's expected non-negotiable dictates! Besides, if a fixed link from Reggio Calabaria on mainland Italy to Messina, Sicily, were completed, say, in the manner of Carlo Alberto Navonne's 1870 AD tunnel (Angelini, 2007, page 39) or a bruited long-distance bridge suspended safely by carbon-nanotubes [fibrous super-ropes], then it should be technically possible to connect Sicily with Sardinia, perhaps to indirectly join all of Switzerland's provinces, traditional and non-traditional! **Figure 5**, Google Image inserted below, illustrates Maurizio Pino and Filomena Francesca Pastore's **Strait of Messina Skyscraper** which won Special Mention at the 2010 Skyscraper Competition held by eVolo. For seismic safety, their skyscraper is duly erected atop reinforced-concrete deeply embedded block-anchorages for both ends of the Messina Strait Bridge's long suspension-cables. Quite literally, the **Strait of Messina Skyscraper** is a death-defying act of macro-planning hubris!

Our Solar System homeworld's first shoreline came into existence when landscape and ocean became distinct bioshell entities sitting atop the flexible, fractured and shifting Earth-crust. The gross form of the Gibraltar Arc is primarily determined by our planet's plate tectonics. However, anthropogenic influences are becoming ever more evident, especially since the management of massive freshwater reservoirs (Scanlon and Smakhtin, 2016) can affect incidentally Earth's natural axial rotation dynamics (Mitrovica et al., 2015). With its first issuance during AD 2016, the new Springer peer-reviewed publication ***Euro-Mediterranean Journal for Environmental Integration*** doubtless has editors uninterested in Macro-Imagineering. Team Geographos proposes the Mediterranean Sea Basin, during the 21st Century, is probably going to become a mecca for the "charm industries", as sagacious Peter John Kaplanis has pointedly instructed Team Geographos on numerous occasions, involving the global tourism market as well as suggested theme-parklands such as R.B. Cathcart's (2015) ***Alexander the Great Sea Dam-Oceanarium*** for Thessaloniki Bay in Greece, designed by a combination of artists and macro-Engineers (Turner, 1998). As a field of endeavor, Sci-Fi Macro-Imagineering is a gedanken-experiment, in which Team Geographos develops a geographical scenario in which most factors are Nature's constants and one anthropogenic factor is permitted to vary so that our job, as disciplined professionals, is to infer logically, sometimes merely reasonably, the geographical results of variation in the significant chosen anthropic factor, resulting in an imaginary world bioshell or defined sub-regions therein! Whilst China [PRC] has rapidly industrialized, that ecosystem-nation's 21st Century leaders are unlikely to assume the aura of a world-class bully by unilaterally commencing global Geo-Engineering of Earth's atmosphere (Moore et al., 2016). Instead, China is methodically extending its geopolitical influence, mainly westwards, through Asia, Russia, Africa and Europe by means of a myriad of strategic mega-projects such as railways, highways, cargo and cruise-liner seaports as well as solidly affixed and flexible hose-type pipelines. Ports in Greece have made deals with China. The Mediterranean Sea Basin might become a tourist destination hotspot charmingly attractive to China's prosperous populace of potential travelers. Imagine our world knit together culturally by coordinated Hyperloop systems before AD 2050! **Figure 6**, below, a Google Image of the Hyperloop One "World Subway Map", which is reminiscent of the "World Metro Map" designed by Mr. Mark Ovenden in 2003 It copies, on a grander scale, London, England's classic layman-understood Underground train routes map first devised during the early-1930s. It was not until AD 1947 that the speed of sound in air, 340 meters per second, was encountered and exceeded by a manned airplane propelled by a rocket engine. It is plausible that oceanic routes might need supersonic submarines, submerged super-cavitating crewed and autonomous or drone vessels, to cross vast ocean distances at high-speed—that is, at greater than the speed of sound in seawater, 1,450 meters per second. Perhaps it can be accomplished by slipping Hyperloop capsules inside a fast submarine's hull so that a physical transfer of passengers and cargoes can be avoided, and thus the time delay reduced to a minimum!

It was the lack of major mineable iron and coal reserves within the Mediterranean Sea Basin that limited the endemic population's industrial development, although not scientific research, to a narrow path; it was the post-World War II inauguration of oil and gas-fields in northern Africa which markedly shifted the paradigms and trajectories of Mediterranean Sea Basin economies. At the present time, with about 46,000 kilometers of coastline available, there is bound to be some significant future economic and cultural connection between an economically resurgent China and the developable Mediterranean Sea Basin's available seashore real-estate! Preservation of the Mediterranean Sea's extant "Aquaterra Incognito" is, therefore, paramount—meaning that the region's forecasted future sea-level rise—which may possibly be glibly over-hyped by radical Green-dominated media or even factually erroneous since global mean sea-level rise really represents the least certain consequence of climate change—ought to be prevented by near-total exclusion of the North Atlantic Ocean, or otherwise severely limited in the infrastructural damages and livelihood riskiness such a one-meter saltwater rise imposes on delicate heritage landscapes, some already in physical disrepair, and its thoroughly hard-working peoples! By the end of the 21st Century, Istanbul, Turkey will probably be the major vulnerable city that suffers the biggest financial losses, followed by Izmir, from the postulated sea-level rise scenario but Marseille in France, Naples and Venice in Italy, Odessa in Ukraine as well as Athens in Greece will also sustain economic and geographical losses their struggling populations can ill-afford to endure (Zerbini et al., 2017). Direct macro-engineering interventions, all of which will be costly enterprises to undertake and successfully complete individually, will be required at important river deltas: the Ebro, Rhone, Po, Nile and Danube (Maselli andTrincardi, 2013; Anthony et al., 2014). As a consequence of the air's supposedly warming regional climate regime, new deep seawater out-flowing over the Strait of Gibraltar's sill will influence the North Atlantic Ocean in terms of salt and heat input whilst modifying the abyssal layer of the Mediterranean Sea. Nature's Mediterranean Sea is nutritionally impoverished compared to other regions of the world-ocean—that is, ordinarily, its euphotic seawater stratum has nutrient concentrations much too low to feed a large endemic biomass [fish et cetera] and, because of the basin's negative seawater balance as well as the resulting currents, the Mediterranean Sea's deepest seawater layer overflow exports major masses of nutrients into the North Atlantic Ocean, making those amounts unavailable for the basin's primary biomass production! The out-flow began 5.33 million years ago after a neck was first formed in the Gibraltar Arc landform that resulted in a modification of the North Atlantic Ocean's thermohaline circulation and climatic changes instigated by contributing the Mediterranean Sea's warm, saline water to northern latitudes (Hernandez-Molina et al., 2014).

21st Century Role a Mediterranean Sea Artwork!

After the pre-cyborgized ATLANTROPA's prolific public-relations campaign ended with Herman Sorgel's death after being maliciously struck by an motor-vehicle on 25 December 1951, the grandiose macro-project disappeared from the world-public's collective mind, quickly fading into the blurry realm of unrecognized history. But, by AD 1997, the irrepressible American geologist Dr. Robert G. Johnson (born 1922-????) unknowingly resuscitated Sorgel's old idea of a massive dam to fill the Strait of Gibraltar (Johnson, 1997), but with different desired goal. That year the American Geophysical Union had 36,000 members and its weekly glossy-paper membership newsletter, ***EOS, Transactions, American Geophysical Union*** still has a worldwide circulation. Dr. Johnson daringly hypothesized that commonly projected increases of the seawater salinity overflowing the sill at the basin's neck could eventually initiate an Ice Age via high-latitude ocean-air circulation changes that encouraged massive re-glaciation of northern Canada. Later, Spanish hydropower consultant Mr. F. Fernandez Damborena, speaking at the ***Hydropower into the Next Century*** conference of 13 dam-building companies at Gmunden, Austria during October 1999, revealed his AD 2017 cost USD$146 billion mega-project vision: a 29.2 kilometer-long rock-fill embankment put in the water-gap now extant between Spain and Morocco, in a major shipping passage 340 meters deep, that would sit, unbreakable and secure, sited on the Camarinal Sill. It would have a centralized mid-barrier powerhouse, for which Damborena gave no estimated electricity output, possibly sited near the Strait of Gibraltar's middle. [Still later, Drs. Wing-le Chan and Tatsuo Motoi (2003) calculated that an anthropogenic seawater-impervious emplaced barrier plunked therein might decrease the meridional over-turning of seawater near Antarctica with significant climate regime changes following.] Finally, Mr. Jim Gower, serving with the Institute of Ocean Sciences, Fisheries and Oceans Canada in Sidney, British Columbia, Canada, by September 2015 offered a revision of Dr. R.G. Johnson's and Mr. F.F. Damborena's compositionally identical rock-fill water-gap blockade scheme (Gower, 2015). In Spain, the engineer Mr. Juan Valle Auguita, representing Presa Puente Estrecho de Gibraltar S.A., offers a beautiful, professionally-made entertaining YouTube video, "The Solution to the rise of sea level, in the Mediterranean basin" [https://youtu.be/9bbFyKE2DWw] supporting a rock-fill dam with a navigable wide central gap spanned by a long-distance railway and highway-carrying suspension bridge.

Devoid of envious harshness, Team Geographos dismisses these fantastically expensive, almost cartoonish, vamped rock-fill dams because such massive piecemeal barrier installations are monetarily infeasible and technically unnecessary! Approximately 70% of humanity's wealth is vested in land and property. [Luckily, by mid-2017, a working group named the International Financial Accounting Standards had substantially completed a document of rules governing valuation, ethics and measurement of property by experts which enables accountants to assist infrastructure builders with a useful "design to cost" macro-planning technique!] Based on Dr. J. Donkers and his esteemed colleagues (1990), it might require a monetary expenditure of at least 2017 USD$ 932,670,000,000 to protect 360,000 kilometers of Earth's valuable and vulnerable coastline against a one-meter future rise in sea-level! Nearly a trillion AD 2017 U.S.A. dollars, or possibly more! Gadzooks! The weighty rock-fill dams blithely contrived by Dr. Johnson, Mr. Damborena, Mr. Gower and Mr. Auguita are actually low-crested structures—enormous berm breakwaters, in fact—for which each person putting forth his salutary suggestion fails to offer any comprehensive relationships, such as between wave climate, habitat change, beach value—in other words, the integrated approach in designing an utterly gigantic coastal protection scheme which incorporates a structure with questionable structural stability considering the proneness to earthquakes of the region. Probable construction macro-problems are left unaddressed. Much smaller berm breakwaters only began to assume modern designs, led by Baird & Associates of Ottawa, Canada, during AD 1983 and much remains to be learned yet by practicing macro-engineers and coastal managers! Essentially, and like seawalls (Jiang et al., 2017), these berm breakwater proposals are simply scientifically stacked stones which must be, whenever emplaced, an absolutely capable design able to sustain every anticipated incoming wave load—even tsunamis, however such seawater wave-forms may be generated inside and outside the Mediterranean Sea Basin!

Before he finally joined the wiser members of what constitutes AD 2018's international Team Geographos, Mr. R.B. Cathcart (in his own mind) had shyly toyed with the phantasmagorical idea of a massive barrier very similar to those three outlined above, but in the form of an unconventional site-specific outdoor public utility sculpture, chiefly used as a Strait of Gibraltar water-gap stopper—soon to be revealed as the ***GSTB*** Macro-project But, rather than dumped rock fill, his hollow stopper was to be composed of poured concrete threaded with strengthening rebar filled by stored materials with some considerable hefty mass. In essence, his Macro-Imagineering idea was a Land Art-styled, tomb-like storage bin enclosing forever chemically solidified carbon dioxide gas and chlorofluorocarbons, both of which are radical Green-vilified greenhouse gases. The bin was intended to become the nightly well-lit key feature of a 24/7/365 themed parkland, the ***Cyborg Atlantropa International Sculpture Park*** (Cathcart, 1998). The barrier was to have more than merely one utility—namely it was to be a stationary aerospace-plane home-base. [On 17 July 1962, Robert Michael White (1924-2010) became the first human to fly to outer space—his X-15 vehicle reached a maximum altitude of 108 kilometers, six times higher than the Concorde airliners retired from public intercontinental flight service in AD 2003—and subsequently land at his flight's place of origin.] Just imagine automated SKYLON aerospace passenger and cargo vehicles, landing and departing on schedule, to achieve and to return from orbiting in near-Earth outer space, returning to its primary maintenance ground station, the paved flattened top of the huge Gibraltar Strait Dam! **Figure** 7, below is a Google Image of SKYLON.

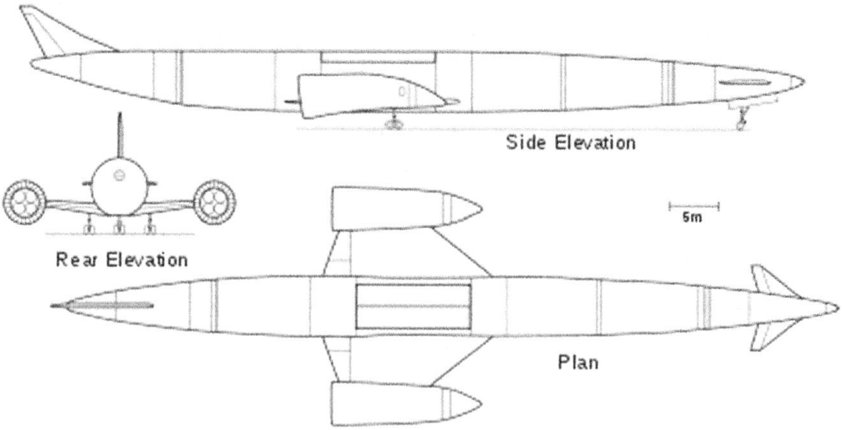

Mr. R.B. Cathcart's fantasized capacious and beautifully illuminated concrete trash-bin, inserted into the Strait of Gibraltar, was to have additional desired economic functions: it was to be a point-deposition place for the outputted solid products of a costly world air Geo-Engineering enterprise—that is, atmosphere control by industrialized direct aerial gas extractions and sequestration. Why such a quaintly drastic macro-project? Because the barrier would generate hydropower on an extremely large-scale! What was the perceived true need for that superfluous barrier-bin mega-project? Two decades ago global warming fears were wildly rampant and Mr. R.B. Cathcart, at that specific time and place, blissfully trusted, rube-like, the radical Green media's unspeakable fear-mongering broadcasts and publications!

The worrisome postulated macro-problem incipient global mean sea-level rise within the Mediterranean Sea region can be effectively, efficiently and inexpensively handled—relative to the four fabulous mounded-stone berm breakwaters briefly referred to above—by another technically feasible and simple means of coping with Nature's unstoppable change through our Earth's continuing, but certainly not unending, Geologic Time. A new Art form, Ocean Art, is hereby announced. Complementing iconic Land Art installations, a team of artist/macro-imagineers proposes the deliberate seawater terracing of the naturally changing Mediterranean Sea and its surrounding landscape. Ocean Art ought to constitute an urgently needed new category of very controllable (post-completion) mitigation Macro-Engineering, a single minutely macro-planned, decorative regional sea-level adjustment solution, seemingly required by the believed prospect of worldwide 21st Century sea-level rise. ***The Mediterranean Sea Terrace Macro-project*** proposal's purpose is to help preserve cherished existing, and irreplaceable artworks, such as Italy's Venice and Turkey's Istanbul that are located on, and perpetually affixed to, the Mediterranean Sea Basin's almost socially magnetic, attractively humanized seacoast. The new art-form, based mainly on the use of seawater-impermeable and strong textiles, requires further technical development a few R&D steps ahead of what has already been accomplished, to date (Povie et al., 2017). All contemporary strategies, and macro-projects, formed to manage the risks imposed by Nature's (or unnatural) future sea-level rise hinge on sound characterizations of all the inherent uncertainties; real-world climate changes will compromise AD 2017's freshwater, materials and energy supply as well as all established and functioning management regimes! When Macro-Imagineering aficionados, lively Sciencceers, contemplate leveling mountain ranges, reclaiming extensive hot deserts, damming wide and fast-flowing rivers and narrow sea-straits with timeless current systems, or inducing regional or global climate regime changes, they should always muse on whether humans have thus revealed themselves to be sightless in the presence of Nature's beauty in doing so or to be suffering from hubris, the overweening species pride or misplaced blind faith in our species' capabilities which can motivate some few megalomaniacal leaders to try to excessively transcend our species' geophysical limitations of the past by undertaking ultra-projects—that is, deliberately contriving ruled group-approved bodacious interferences with the Earth-bioshell's God-given natural systems done in the service, of course, of a perceived Homo sapiens future-realized good. [A beneficial mega-project, were if perfected, may be Dr. David Harrison's. Working at the University of Sydney, he is investigating the practicality of artificial cloud-brightening to prevent coral bleaching of Australia's famed Great Barrier Reef.] Team Geographos finds it puzzlingly odd when expert urban planners who are intellectually anticipating the birth of our planet's next billion persons, ignore or reject both the Sahara and the Mediterranean Sea Basin as "suitable places" for the improvement of extant cities or the establishment of future human settlements (Forman and Wu, 2016)! Simply stated, it is a proven geographical fact that our world's maritime domain is trending to become, amazingly soon, a totally "Cyber Ocean" Nature-Human Complex (Dellios et al., 2015), including a mixture of robotized and autonomous ships, automated seaport management-operation-maintenance along with constant monitoring of weather, sea-currents and infrastructures by swarms of various types of liberated robots, even nearly independent AI robots!

Precursor Art-forms, Ocean Art Progenitors

Space Art proponents have mostly opted to construct various symbolic artifacts in extra-terrestrial space, some of which could be visible from Earth's surface with the naked human eye. Some actual artworks, or proposed artworks, are connotatively indescribable or commemorative constructs. Using various plastic film and dense textile envelopes, Air Art's advocates have already exploited some of the possibilities of compressed air (or other safe-to-use pressure- blown gases extracted from air or ground) as well as Nature's winds. Finished projects of Land Art usually result from different personal interpretations of natural and anthropogenic landscape significance and their alteration by artists and others. Artworks were installed in the world-ocean near Tobago, West Indies during AD 1969 by Peter Hutchinson and Dennis Oppenheim and, later, during 1983, Christo installed eleven flamingo-pink floating plastic collar-mats, covering approximately 600,000 square meters of a Biscayne Bay lagoon in the State of Florida, U.S.A. Installations of this type were first dubbed "Oceanographic Art"—they are best apprehended as decorations. Ocean Art, however, is a form of seawater sculpting by aquatic terracing focused on the +70% of our Earth's area that is the world-ocean. Seemingly, and perhaps actually, the originating conceptualizer of Ocean Art is the renowned freelance German architect Frei Otto (1925-2015), **Figure 8**, below, a Google Image, the titan of 20th Century tent architecture who first elucidated his structural conceptualization circa 1953 AD, a year after pre-cyborg ATLANTROPA-devisor Herman Sorgel's untimely demise. In Team Geographos' view, Ocean Art has practical, commercially-useful aspects associated with that descriptive technique label that "Oceanographic Art" lacks and, therefore, is of interest to 21st Century professionals of Macro-Imagineering and Macro-Engineering! With regard to the Mediterranean Sea Basin, Team Geographos refers strictly to 21st Century professionalized **Sea Art** (Pugno et al., 2011), an uncompromising spin-off from Dr. Martin Newberry Cullen's innovation, because such a name particularizes the macro-planning focus, and long-term impact, of our geographically large-scale region design for a popular world traffic crossroads oceanic segment.

Staircase Farming: Humankind's Lasting Ancient Land (-scaping) Art Effort

About 5,000 years ago, the Earth-atmosphere's methane concentration increased markedly and its main source then was Homo sapiens' cultivation of rice crops on flatland paddies; about 2,000 years ago, humans boldly commenced growing rice in watery paddy terraces etched laboriously into hillsides. Terracing refers to the bench, also known as a terrace landform, constructed in the shape of a ridge and channel the entire earthen soil surface of which is cultivated as a tended farm-field. For example, the spectacular Philippines tourist attraction of awe-inspiring hillside rice paddy terrain at Ifugao, northwest Luzon, is preserved as a UNESCO World Cultural Heritage site. UNESCO's dedication reflects people's unwillingness to accept the natural duration of their grandest outdoor physical public creations, Art or otherwise, since, currently, a global financial undertaking helps to prevent natural erosion as well as unintentional or deliberate human actions causing obvious visible decay, and ultimately possibly the erasure of this very ancient and entrancingly picturesque anthropogenic terrain. [Drs. Leoncia Calayag Roxas and Samuel Medina-Roxas, including all members of their friendly, well-coached team of proficient repairers, have generously given Team Geographos extra-sufficient helpful geographical data, as well as informative personal experience commentary, on this specific real-estate landscaping subject-matter.]

Methane is a greenhouse gas and the anthropogenic contribution to the Earth's air causes some enhanced global atmospheric warming and, consequently, may also contribute to the slight, but ongoing, rise worldwide of the world-ocean's level (Rigby et al., 2017). Perhaps half of all living humans eat rice and the micro-organisms inhabiting anoxic farmland rice-field soils contribute between 10% and 25% of Earth's annual gaseous methane emissions; by AD 2030, there might be more than five billion persons consuming rice regularly. Artificial wetlands on shaped hillsides have in the past, and continue still to contribute the world-ocean's instability in terms of volume and area; by AD 2100, our homeworld's ocean could sit one-meter higher relative to its AD 2017 level, thereby directly affecting the length and features of our planet's accurately mapped, inaccurately mapped, unknown or simply unmeasured coastline. Worst-case global warming geophysical scenarios focused on the possible permanent submergence of some of Earth's landmass have been assembled for display by famed artists such as Peter Fend (1999), Terry Schoonhoven as well as Helen Mayer Harrison and Newton Harrison (Heartney, 2003). Quarry Cove, a part of the Yaquina Head Outstanding Natural Area near Newport, State of Oregon, by 1994 served to permanently remove a minute volume of the world-ocean's seawater—insufficiently capable of any anti-global sea-level rise macro-engineering management mega-project (Newman and Fairbridge, 1986). [Quarry Cove's minuscule beneficial reduction was easily displaced upwards by Al Nakheel Properties' continuing anthropogenic island construction that extends Dubai's natural post-Ice Age sandy shoreline farther seaward.]

Mediterranean Sea Terracing: 21st Century Appraisal by Team Geographos

About 4,300 years ago, urban governments commenced construction of edificial monuments and massive infrastructures following the post-Ice Age fleeting stabilization of Earth's ocean. About 2.2% of this planet's landmass that is ten meters or less in aerial topographic elevation probably supports +10% of all humans, of which about +13% are classified as "urbanized". Humanity's collective activities to make and earn a living will, inevitably, globalize the Mediterranean Sea Basin—its seawater, organic and inorganic contents, and highly incidented periphery. Mediterranean Sea Basin ecosystem-nations can utilize pre-21st Century technologies to avert a costly permanent submergence by seawater. However, recent R&D and newer industrial products derived from advanced material technologies—particularly, technical textiles and films exhibiting high-performance, purely functional, and precisely woven on cast fabrics—offers teamed artists/macro-engineers the prospect of a cheap Mediterranean Sea Basin hung safeguard, an adjustable and, if needs be, removable anti-sea level rise barrier placed underwater in the Strait of Gibraltar. A fabric artwork and dam, the Gibraltar Strait Textile Barrage (***GSTB***), could supplement the AD 2018-completed Modulo Sperimentale Elettromeccanico (acronym: MOSE) installed to "save" universally revered Venice, Italy, by costly installation of an operationally difficult-and-to-manufacture-and-maintain complex of lifting storm-surge gates that will be timely raised to secure the fragile city's surrounding lagoon (Laborataoir Bale, 2017). Modern-day Venice has exceeded its carrying capacity, a crisis hastened by the onset of enormous cruise-liner visits, veritable floating enclaves of booze and entertainment-pacified ambling tourist day-visitors (Weaver, 2005). Team Geographos proposes that Venice's authorities take notice of the artist Christo's most recent public artwork, his 2016 AD ***Floating Piers*** on Lake Iseo in northern Italy, nestled amongst the imposing foothills found at the base of the Alps. During 2016 AD, it was the world's most visited artwork, with an average of 75,000 persons daily walking on placid naturally pooled freshwater! Christo made it possible for transient visitors to ramble safely 50 centimeters above Lake Iseo's surface, from the lakeside town of Sulzano to Monte Isola, on his 3 kilometer-long, buoyant, bright-yellow linear fabric swatches of a walkway affixed to 220,000 high-density, foamed polyethylene floatation cubes roped to artfully placed heavy concrete anchors resting 90 meters below upon the lake's muddy bed. Team Geographos asks, meekly: why could not such a beautiful, non-energy-using means of pedestrian pleasure replace at Venice the flitting smog-generating motor-launch taxis and ponderous diesel exhaust-puffing cruise-liners, and their superfluous ugly anchorages and necessary mooring appurtenances? Possibly the ugly railroad bridge to Venice could be demolished after floating walkways are installed. **Figure 9**, below, a Google Image of bridge first built in 1846. What an ugly, blunt pointy intrusion: rains, cars and enveloping smelly smog.

Of course, gondolas ought always in future to have super-priority over all other means of moving people in Venice! The prospect of pedestrians, walking on seawater, has a certain fascination in the mind's eye. Imagine if the fabric walkway were decorated, in the ornate and beloved Venetian style, and even periodically changed-out to provide tourists who visit Venice more than once with a new, dramatic and elegant stage-setting! Under some climate change modelling, the Mediterranean Sea Basin becomes uncomfortably warm during summertime, but a more pleasant destination in Spring and Autumn (Amelung and Viner, 2006); here again, an immediate R&D and factory investment opportunity arises: Italian mill-woven fabrics and Italian-designed clothing styles are admired and worn by millions of people and, since the region is probably going to become warmer, why not adapt clothing that cools the wearer by 2^0 to 3^0 Celsius better than the best cotton garments (Hsu et al., 2017) and exudes characteristic Italian quality and elegance of styling? Indeed, rather soon this century a huge non-luxury clothing market for such wearables will be opened by the then extant Arabo-Persian Gulf's heat-besieged suffering populace!

Obviously, the Strait of Gibraltar connects the North Atlantic Ocean and the Mediterranean Sea, making it inevitable that the Mediterranean Sea could rise as our planet's world-ocean elevates, for whatever reason of Nature or Homo sapiens. Team Geographos' ***GSTB*** (Gibraltar Strait Textile Barrage) will likely be draped, barring interminable international legal disputations and intransigent geopolitical objections (Caminos and Cogliati-Bantz, 2014; Martin, 2014; Linnell et al., 2016), on a 20 kilometer-long alignment between Tarifa in Spain and Ksar e'Sghir in Morocco, creating an aerial and submarine fabric artwork somewhat imitative of Christo's ***Valley Curtain, Rifle, Colorado, 1970-1972***, which was macro-engineered by Ernest C. Harris (1915-1998). Its sole purpose will be to insure the maintenance, for a long period of future Geologic Time, the present-day Mediterranean Sea's level or possibly to lower it enough to significantly reduce the existing stresses and strains on the Basin's Nature-degradable infrastructure. In other words, this immense artwork ought to preclude all future erosion of valued land-sited artworks encompassed by the urban fabric of northern Africa, parts of the Middle East and southern Europe! On a small scale, the ***GSTB***'s structural, mechanical and hydro-dynamical Physics was inspiringly first hinted by demonstrated mathematical calculation during AD 1974 by Dr. Andrew Noel Schofield (born 1930): U.S.A. Patent 3,785,158, "Hydraulic Engineering Installations", awarded 15 January 1974 and during AD 2005 by U.S.A. Patent Application Publication US2005/0163570 A1, "Flexible Water Gate", issued 28 July 2005 to Dr. Martin Newberry Cullen (born 1956), both eminent UK professional engineering men. For more than half a century geomembranes have functioned successfully as a means of underwater sealing (waterproofing upstream faces of artificial water barriers) of large dams on land. By AD 2016, fiber-reinforced polymer lock-gates, the largest in the world, were installed and functioning properly as expected in the Wilhelminakanaal in Tilburg, The Netherlands. (SEE also: Bugaud et al., 2000.)

Valley Curtain, Rifle, Colorado, 1970-1972 was the notional forerunner of Team Geographos' ***GSTB.*** (Christo's installation artwork was punctured to prevent its being torn asunder by strong up and down valley wind gusts.) Coincidentally, during 1971-1972, a watery version similar to Christo's was concocted by UK engineer A.N. Schofield as a competition proposal for the then prospective, and later realized, Thames River Storm Surge Barrier mega-project. The total area of the more or less vertically hung ***GSTB*** is about 200 square kilometers but only approximately 20,000 square meters of the drape's eastern bowed facade will be exposed, simultaneously, to one-meter head of North Atlantic Ocean seawater on its western face, ambient air and material-degrading sunshine on its eastern face. Team Geographos foresees adjustments being made by shore-based winches, lifting the ***GSTB*** in gradual increments if the world-ocean does as expected during the 21st Century—that is, experience sea-level rise. In other words the ***GSTB*** mechanism performs a plastic form-active structural act of compensation to accommodate Nature's changes. The Sea's evaporation losses generate a seawater deficit that is compensated by a net inflow of North Atlantic Ocean seawater, which is progressively transformed within the Sea's basin, becoming saltier and eventually descending to intermediate and deep strata in convection processes triggered by wintertime cooling. Finally, the colder and saltier seawater leave the basin through the constricted unblocked Strait of Gibraltar (Borghini et al., 2014). In the unlikely event-process the drape should snap from overstrain whilst withholding a one-meter North Atlantic Ocean seawater head, then a roiled inrushing seawater-wave, at first curving slightly to the southeast because of Earth's rotation, will unstoppably spread throughout the Mediterranean Sea Basin. However, this flood—subsequently, the ***GSTB*** breach may be countered by simple re-erection of the ***GSTB***, and allowance of Nature's normal aerial evaporation of seawater within the Basin, a loss of about 50 to 100 centimeters per year—cannot ever be of such tumultuous violence that it imprints cultures lodged within the Basin with their own special vocabulary of a 21st Century flood Ur-myth!

Dr. Martin B. Widden, since circa 2001 AD, has developed "Siphonic Hydropower" systems at Lancaster University in the UK. If the available seawater head is only one-meter at the working ***GSTB***, the losses in the Siphonic Hydropower system (aeration, due to pipe friction and due to upward drift of the air bubbles) would use up probably 0.4 meters of the available North Atlantic Ocean seawater head. Thus, the efficiency of the system should be about 60%, without considering losses in the air-turbines and losses in electric cables. A ***GSTB*** Siphonic Hydropower system of the size that might be labeled an Ur-myth flooding [106.000 cubic meters per second] would justify the installation of very efficient air turbines, at least 80%, so that overall efficiency could be 50%. Possibly, under the very worst of times and conditions, the electricity output of the ***GSTB*** can be as much as 500 Megawatts. In a real-world non-catastrophic progressive scenario, the actual steady output would actually be much less. One especially good use to which any ***GSTB***-generated and transmitted Siphonic Hydropower may be put is the powering of far-distant wind power-driven air bubblers [aerators] (Stigebrandt, 2016) situated (as in room-sized aquaria!) at

Mediterranean Sea hypoxic regions (coast of southern France, the coast of Greece and the northern Adriatic) since after the ***GSTB*** installation's it has become the world's largest oceanarium—wherein valuable fish face shortened lifetimes, or none at all (Piroddi et al., 2017). For sure, future climate regime changes could increase Basin-wide air temperatures and decrease precipitation directly on, or draining into the Mediterranean Sea via rivers, thus causing substantial thermohaline seawater circulation alterations (Powley et al., 2016). At the present time, within the Mediterranean Sea Basin, only the ecosystem-nation of France has the bio-capacity to be self-sufficient in terms of food-stocks (Galli et al., 2017).

Speculative Macro-Projects, Cyborg Adriatic Sea

Competitive ecosystem-nations centered on the northern half of the Adriatic Sea require speculative macro-projects that entail public-sector investment in, and often prolonged geopolitical negotiation of, large-scale coastal and transportation infrastructure necessary to achieve economic development objectives. Such mega-projects are speculative because there is never a guarantee possible—that is, if you build it, will [say, Chinese] people expend money to visit! Team Geographos proposes a special kind of "industrial region", one warranted to the process-event of international tourism within the northern half of the Adriatic Sea and its adjacent strand, as well as its wider landscape, the northern Adriatic Sea-bordering coast. This suggested region might be termed a "Gateway" to further geographic aggregation and dispersal of freight and human passenger intermodal flows to inland and overseas destinations that are potentially far from Venice, Italy. [GOTO: www.tirol-adria.com for details of an Alps Canal Tunnel Waterway.] Sixteenth Century Venetians celebrated the Earth-bioshell's potentially compleat habitability because of their technologically newly-won navigability of the world-ocean (Headley, 1997). Climate regime change is expected to alter the global mean sea-level of the tidal world-ocean somewhat so it has been bruited by Adam Lowe and Jerry Button (2015) that a non-swimming "Olympic" pool-size temporary machine, 20 meters wide by 40 meters long by 3 meters deep, be constructed on the Island of San Giorgio Maggiore inside the Venice lagoon so that, when the ***Terra-Forming: Engineering the Sublime*** is gradually filled with pumped seawater, its slow filling will clearly demonstrate to live-audiences present, as well and Internet viewers, exactly what occurs to all of our planet's seashore landscapes if ever the Earth's future world-ocean increased its seawater volume substantially! Such a three-dimensional toy of sea-level rise sited on the Island of San Giorgio Maggiore should be extended visually to the underground wherein seawater can be injected by wind-power pumps to help expand deep city-supporting aquifers thus raising the seafloor beneath subsiding Venice, Italy (Gambolati and Teatini, 2014)! Wind-powered seawater injection by floating wind turbines are slated to be tested by ExxonMobil, Nexen Petroleum UK, and Statoil amongst others.

Since aquifers have a finite capacity to hold saltwater, when they are full then the turbines will produce power for people. U.S.A. designers Yitan Sun and Jianshi Wu won First Place in the eVolo 2016 Skyscraper Competition with their ***New York Horizon***. Accurately described, it is simply a continuous uniform height mirrored skyscraper edging the perimeter of New York City's magnificent Central Park after it had been Green-restored to its primeval, Ice Age-stripped rocky state (Steinberg, 2014; Sanderson, 2009)! ***New York Horizon*** is an architectural folly as well as a rip-off of Italy's 1960s **SuperStudio** architectural design concepts! Heaven forbid intellectually calloused macro-imagineers ever treat Venice so cavalierly!

Certainly the cyborg maritime realm, a sub-domain of the larger critical infrastructure cyborg realm, will play a decisive future role in the worldwide industry of tourism. Global trade is embracing highly automated systems and machinery, with minimal human intervention, to handle the mountainous volume of goods that international agreements bring to the docks of seaport cities. Drones are now widely used to shuttle mail, medicine, spare parts and to inspect hulls. Drones are being designed to load and unload small cargo containers quickly from ship to shore and vice versa! Swarms of underwater robots, perhaps numbering in the thousands or millions of simultaneously operating units, could potentially be deployed to capture real-time three-dimensional oceanographic data on the Adriatic Sea's condition (Jaffe et al., 2017)—that is, to provide accurate assessments of its usefulness! The northern Adriatic Sea is subject to anoxia, harmful algal blooms and mucilage concentrations. Team Geographos geo-visualizes (Newell and Canessa, 2017) a robot-monitored 21st Century world, populated by self-buried, drifting lighter-than-air and drifting Mediterranean Sea-immersed sensor devices, generally disguised to not look remarkably out-of-place anywhere in Nature, that will constantly signal the status of operation of both the planet and human civilization's networked infrastructures—that is, a computational fabric totally covering our planet that may, ultimately, become sentient. Linked into the cyborg domain will be all future amphibious buildings wherever installed close to the cyborg Adriatic Sea—that is, buildings (homes, factories et cetera) that sit on ground during yearly periods of normal weather conditions but can rest afloat, with equal structural stability, whenever rising seawater overspills by intrusion on the foundational ground (Zevenbergen, 2017). AD 2016 was the 50th Anniversary of Venice's l'Acqua Granda of the 4th of November caused by a peak storm surge that arrived during low tide (Trincardi et al., 2016).

A cross-Adriatic Viaduct, spanning the body of seawater from Zadar (Croatia) to Ancona (Italy) was proposed during AD 2009 by Mr. Giorgio de Romanis (born circa 1937?), **Figure 10**, below, in his extraordinary macro-imagineering text ***Il Ponte Sull'Adriatico: The Bridge Over the Adriatic***. (GOTO: http://www.ilpontesulladriatico.com.)

Our world's longest sea-bridge, the 29.6 kilometer-long Hong Kong-Zhuhai-Macao Bridge, opened in China during 2016 AD. The northern Adriatic Sea viaduct may be safely located in shallow seawater. **Figure 11**, below, a Google Image, indicates the depth is 60 meters or less.

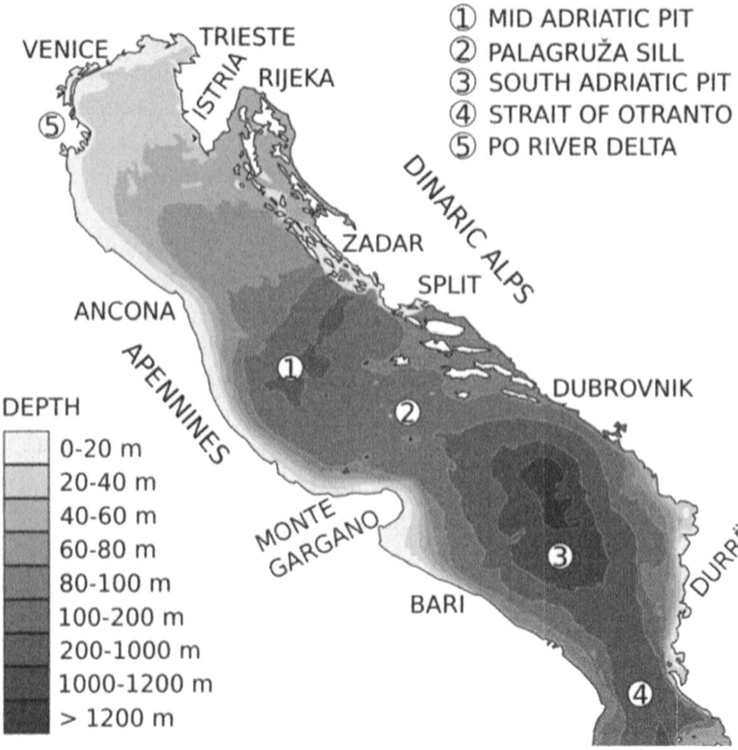

Tsunamis have been documented in the Adriatic Sea at least since the 16th Century—some caused by meteorological process-events, such as the Bora, a northeast winter wind sometimes of gale force coming from the eastern-side of the Adriatic Sea, others from unknown or surmised causes (Sepic et al., 2012; Pasaric et al., 2012). Perhaps instead of a sea-bridge, such aerial and oceanographic process-events might be uncompromisingly endured better by a Baltic Sea submarine Hyperloop installation such as that which has been planned since AD 2016 to connect Turku, Finland with the Aland Islands and thence to Norrtalje, Sweden.

Although Mont Saint Michel, France, became an artwork proudly produced by humans during a short period of Geologic Time, by both famous and nameless macro-engineers, it was never a place intended to be industrialized (Lefeuvre and Bouchard, 2002). During the 20th Century, anthropogenic islands on which to situate industry were proposed for shallow places offshore England and the Netherlands (Marsh, 1977) as well as Belgium (Charlier and de Meyer, 1992). By resemblance, vast offshore windfarms have been macro-planned since circa AD 2001 for Europe's coastal waters—the anticipated European Offshore Super-grid—which includes a subsea electricity grid linking offshore windfarms. At Brussels, Belgium on 23 March 2017, TenneT TSO B.V., Energinet.dk and TenneT TSO GmbH signed an agreement to create, on Dogger Bank in the North Sea, an artificial island-based wind power hub, **Power Link Island**, with cables connected to at least four ecosystem-nations. At an altitude above the ocean, typically about 500 meters, the influence of the ocean surface's friction on ambient air-flow is diminished greatly; offshore wind generators are best located in seawater less than 50 meters deep to allow bottom-mounted turbines and to be nearest to consuming urbanized landscape zones (Sun et al., 2012). Offshore wind-farms consist of an array of individual wind turbines that each capture the kinetic energy from wind and convert it to electricity; since the electricity is generated in a widely-distributed manner, it must be transmitted through submarine HVDC cables to controlling hubs where it is there collected, conditioned and transmitted to AC-using consumers on land. So far, the European Offshore Super-grid does not include any installations in the Adriatic Sea, only some installations east of Spain, south of France in the Mediterranean Sea intended to feed power into two separated national grids. In the Adriatic Sea, offshore windfarms might best be sited north of Mr. Giorgio de Romanis' lengthy sea-bridge or a seabed tunneled, or an anchored immersed floating tube, Hyperloop route. These offshore windfarms (Schweizer et al., 2016) should send submarine cabled Green electricity directly to Venice, which then ought to become a mid-21st Century controlling hub for the Adriatic Sea Electric Grid system serving onshore regions bordering the northernmost Adriatic Sea.

Since the orbiting ball-shaped shiny **Sputnik I** sparkled Homo sapiens' Space Age during 1957 AD, and the post-1990 AD Internet connects with a myriad of Earth-orbiting active satellites, it seems wise to Team Geographos' membership that humans adopt a vertical, instead of horizontal, outlook on natural and man-made islands surrounded by seawater for the simple reason that, besides being bounded by seawater, islands are also actually defined by our Earth-crust's variable-depth Moho—the Earth's crust-mantle boundary—below as well as by outer space 100 kilometers above! Indeed, we endorse the architectural concept that islands should be conceived as a kind of Architecture autonomous from these boundaries. Kimberly Michelle Sutter's U.S.A. Patent 7,036,449 B2, "Man-made island resort complex with surface and underwater entertainment, educational and lodging facilities", awarded 2 May 2006 truly exemplifies and consolidates our Team's general thinking on this facet of Macro-Imagineering.

Our Earth-bioshell's Terminus Technicus?

Before its far-future deliberate anthropogenic disassembly, our planet's Geologic Time may reach its final chronostratigraphic state when its bioshell—earlier merely an emergent super-organism with its own teleology, desires and needs (Haff, 2014)—becomes transmogrified into a "Bio-Adapter", an unprecedentedly voluminous biocybernetic garment. Just as a spacesuit is a back-up for a spacecraft, the enchanted "Bio-Adapter" is a back-up for a changing Earth-bioshell. And, literally, a "Bio-Adapter" can be haute couture! Homo sapiens (De Preester, 2011; Liao et al., 2012) must ultimately become the main force causing Earth's change into the first planet-sized Bio-Adapter—that is we, the people, are the only known species in our Solar System capable of architecturally designing the selective, complicated conversion of Nature's most desirable planets into life-maintaining machine-entities. [Obviously, Team Geographos' assertion contradicts the outlook of those who have imagined the future Earth, after the end of its ceno-Anthropocene, totally without a population of human beings (Weisman, 2007; Zalasiewicz, 2008)!] So, the question Team Geographos must satisfactorily answer for our readers must be: "What are Bio-Adapters?" (Tabbert, 2005; Pirni, 2016). Answer: a synonym for "Gaia" (Nerlich and Jaspal, 2012), since Homo sapiens seems to be evolving towards a single hive-mind because of the existence of "...networking technologies specifically designed...[by us]...to enhance our networking prowess..." and penultimate success (Coward, 2016, page 86)! Koert van Mensvoort and Hendrik-Jan Grievink (2011, page 460-461) have revised Arthur C. Clarke's earlier declarative remark—SEE **Preface** to this text—by their announcement: "Any sufficiently advanced technology is indistinguishable from Nature". One of our senior teammates, Dr. Alexander A. Bolonkin, has delved into the difficulties of sending a human being into the hard vacuum of outer space without a spacesuit and found a means to triumph over cosmic Nature (Bolonkin, 2009)! Ejection into space without a protective garment presents physiological stresses that humans are not naturally composed by Nature to overcome (Homer,

2017). Yet, a human would not instantly freeze because heat transfer from the body to a space vacuum is slow; the absence of normal atmospheric pressure can cause ebullism due to the formation of water-vapor under the skin that leads to embolism, the blockage of blood vessels by gas bubbles. However, hypoxia [oxygen deprivation] is especially plaguing since oxygen diffuses out of the human bloodstream when the lungs are exposed to hard vacuum. And, deoxygenated blood heart-pumped to the brain then instigates unconsciousness, followed shortly by death.

During AD 1962 the Austrian gentleman, Mr. Oswald Wiener (born 1935), **Figure 12**, below, a Google Image, began writing a book that, eventually, elaborated the amazing consequences were our Earth-world replaced by sequestered humans (Wiener, 2001). Although the first model of Wiener's postulated Bio-Adapter would be a single external—extra-somatic—entity, at the end of Geologic Time, it would still be a single transformational external entity—but holding a consumed planet instead of only one digested human being, possibly a lone Novo ultra sapiens sapiens! Imagine the Austrian artist-philosopher Mr. Oswald Wiener's prototype Bio-Adapter as a more visceral, active extension of the room-sized "ultimate display" forecast by Ivan Edward Sutherland (born 1938), that Team Geographos prominently featured in Chapter One of RB. Cathcart's ***Sci-Fi Macro-Imagineering*** (2017; SEE also: Balint and Hall, 2015). Enclosure and consumption of one Homo sapiens person results in a basic ergoshell. An aggregation of operational and interconnected ergoshells that enshroud a generally spherical celestial body, such as Earth and Mars, results in a biocybernetic planetary "Bio-Adapter Shell". [Today's Earthly bio-shell is just an incomplete rudimentary precursor requiring intensive future biogeochemical and geophysical adjustment by our civilization. "Consequently, if civilization as a whole has any kind of future is not definitely anchored to the Earth, but by traveling among the stars..." (Rodriguez et al., 2017, page 50).] It follows, logically, that Mars and Earth might co-exist someday as separate technology-revamped worlds with "Bio-Adapter" shells added—we might call each of them a "Gaia"—but even these may eventually be incorporated within a star-enclosing, big swarm-formed biocybernetic Dyson Shell fashioned from our Solar System's pulverized and repurposed materials (Cathcart, 1983). Team Geographos has no idea of what kind, if any, "Bio-Adapter" archaeological techno-signature might remain after extinction of Homo sapiens or extinction/absence of any long-ago Alien species Solar System visitors!

In essence, the enchanted Bio-Adapter, whatever the geographical or astronomical spatial scale, can be presumed a namesake variant of a profound Homo sapiens, or post-Homo sapiens invasion ecology strategy that is functionally super-efficient (Dick et al., 2017). "Strategy" is meant as an unrestrained specification of the critical and consequential positions of the subject. Mr. Oswald Wiener alleged that the purpose of his "Bio-Adapter" was to supersede the Earth-world, first by replacing an obsolete, indeed rather crudely mothering, Nature. Is it possible that his perfected garment, enlarged to dimensions too big even for Team Geographos to macro-imagine, and which certainly cannot be compared to cannibalism or mundane planned obsolescence, could become our Universe's ultimate cyborg-gadget allowing "us" to enter, traverse and pass beyond the Cosmic Micro-Wave Background veil into a different, pristine, fertile and convertible Universe? Mr. Oswald Wiener's avant-gardeness is one that totally squelches all imaginable succeeding pseudo avant-gardes—the perfect true Green-desired, but strictly anthropogenic, Universe and extra-Universe 100% recycling scheme!

References Cited

Abadie, L.M. et al. (16 December 2016) Climate Risk Assessment under Uncertainty: An Application to Main European Coastal Cities. *Frontiers in Marine Science* 3:1-13.

Abdel-Fattah, A. and Pingitore, N.E. (2009) Low levels of toxic elements in Dead Sea black mud and mud-derived cosmetic products. *Environmental Geochemistry Health* 31: 487-492.

Abdullah, A.D. et al. (2016) Predicting the salt water intrusion in the Shatt al-Arab estuary using an analytical approach. *Hydrology and Earth System Sciences* 20: 4031-4042.

Abotalib, A.Z. et al. (2016) Groundwater processes in Saharan Africa: Implications for landscape evolution in arid environments. *Earth-Science Reviews* 156: 108-136.

Abril, J.M. and Perianez, R. (1 April 2017) A modelling study on tsunami propagation in the Red Sea, potential hazards and spectral analysis. *Ocean Engineering* 134: 1-12.

Aggestam, K. and Sundell, A. (15 April 2016) Depolitcising water conflict: functional peacebuilding in the Red Sea-Dead Sea water conveyance project. *Hydrological Sciences Journal* 61: 1302-1312.

Albanese, C. and Letzing, J. (2 June 2015) Thinking of a Mediterranean Vacation, How About Switzerland? *The Wall Street Journal* CCLXV, #127: A1 and A10.

Alexander, L.M. (April 1992) The Role of Choke Points in the Ocean Context. *GeoJournal* 26: 503-509.

Almalki, K.A. et al. (August 2015) The Red Sea—50 years of geological and geophysical research. *Earth-Science Reviews* 147: 109-140.

Al-Omari, A. et al. (2014) The Red Dead Canal project: an adaptation option to climate change in Jordan. *Desalination and Water Treatment* 52: 2833-2840.

Amelung, B. and Viner, D. (2006) Mediterranean Tourism: Exploring the Future with the Tourism Climatic Index. *Journal of Sustainable Tourism* 14: 349.

Anatharaman, H. et al. (August 2015) Dynamic properties of silicon carbide hollow particle filled magnesium alloy (AZ91D) matrix syntactic foams. *International Journal of Impact Engineering* 82: 14-24.

Angelini, A. (2007) *Il mitico Ponte sullo Stretto di Messina.* Milano: FrancoAngeli s.r.l.. 286 pages.

Anon. (May 1929) No title. *Railway Journal* 35: 34. [SEE also: "A Subway through the Sahara", *Modern Mechanix: Yesterday's Tomorrow Today* (September 1929). Page 52.]

Anon. (1999) Bridging the Strait of Gibraltar. *Transmission & Distribution World* 51: 18.

Anthony, E.J. et al. (December 2014) Human influence and the changing geomorphology of Mediterranean deltas and coasts over the last 6000 years: From progradation to destruction phase? *Earth-Science Reviews* 139: 336-361.

Anzidei, M. et al. (2017) Flooding scenarios due to land subsidence and sea-level rise: a case study for Lipari Island (Italy). *Terra Nova* 29(1).

Augustin, N. et al. (1 December 2016) Geomorphology of the central Red Sea Rift: Determining spreading processes. *Geomorphology* 274: 162-179.

Azarenkov, N.A. et al. (2014) Hydrogen sulfide energy of the Black Sea. *East European Journal of Physics* 1: 4-21.

Badescu, V. et al. (2008) Sand Dune Fixation: A Solar-Powered Sahara Seawater Pipeline Macroproject. *Land Degradation & Development* 19: 676-691.

Badescu, V. (March 2009) Asteroid impacts or nuclear explosions in the northern part of Black Sea and risks for poisoning the inland population. *Journal of Environmental Assessment Policy and Management* 11: 131-159.

Badescu, V. et al. (2010) Transatlantic Freshwater Aqueduct. *Water Resources Management* 24: 1645-1675.

Badescu, V. et al. (2010) Tsunamis and Poisonous Gases Generated by Asteroid Impact in the Black Sea. Pages 1-28 ***IN*** Veitch, N. and Jaffray, G. (Eds.) *Tsunamis: Causes, Characteristics, Warnings and Protection.* New York: NOVA Publishers.

Badescu, V. et al. (2013) Desalination Brine Disposal by Submerged Pipes in the Red Sea. *Journal of Coastal Research* 291: 81-92.

Balint, T.S. and Hall, A. (May-June 2015) Humanly space objects—Perception and connection with the observer. *Acta Astronautica* 110: 129-144.

Bashitialshaaer, R. and Persson, K.M. (2011) The Dead Sea Future Elevation. *International Journal of Sustainable Water and Environmental Systems* 2: 67-76.

Behnke, R. and Mortimore, M. (Eds.) (2016) *The End of Desertification? Disputing Environmental Change in the Drylands.* The Netherlands: Springer. 560 pages.

Beshkov, V. et al. (2015) Energy Production from Black Sea Water by Sulfide-Driven Fuel Cell. *15th International Multidisciplinary Scientific Geoconference SGEM, Energy and Clean Technologies CONFERENCE PROCEEDINGS.* Pages 207-213.

Beyth, M. (August 2007) The Red Sea and the Mediterranean-Dead Sea canal project. *Desalination* 214: 364-370.

Bikoulis, P. (2015) Evaluating the impact of Black Sea flooding on the Neolithic of northern Turkey. *World Archaeology* 47: 756-775.

Boivin, N.L. et al. (7 June 2016) Ecological consequences of human niche construction: Examining long-term anthropogenic shaping of global species distribution. *Proceedings of the National Academy of Sciences* 113: 6388-6396.

Bolonkin, A.A. (2009) Man in Outer Space Without a Special Space Suit. *American Journal of Engineering and Applied Sciences* 2(4): 573-579.

Bondesan, A. et al. (2013) Geomorphology of the El Alemein Battlefield (Southern Front, Egypt). *Journal of Maps* 9: 532-541.

Book, J.W. et al. (June 2014) The oceanic response of the Turkish Straits System to an extreme drop in atmospheric pressure. *Journal of Geophysical Research: Oceans* 119: 3629-3544.

Boos, W.R. and Korty, R.L. (2016) Regional energy budget control of the intertropical convergence zone and application to mid-Holocene rainfall. *Nature Geoscience* 9:892.

Borghini, M. et al. (2014) The Mediterranean is becoming saltier. *Ocean Science* 10: 693-700.

Bowman, D.M.J.S. et al. (2017) Human exposure and sensitivity to globally extreme wildfire events. *Nature Ecology & Evolution* 1: 0058.

Boyle, M.O. (March 2008) Harvey in the sluice: from hydraulic engineering to human physiology. *History and Technology* 24: 1-22.

Brock, R. et al. (1999) A case study of efficacy of freshwater immersion in controlling introduction of alien marine fouling communities. *Pacific Science* 53: 223.

Brothers, D. et al. (2011) Loading of the San Andreas fault by flood-induced rupture of faults beneath the Salton Sea. *Nature Geoscience* 4: 486-492.

Bugaud, M. et al. (2000) Health monitoring of composite plastic waterworks lock gates using in-fibre Bragg grating sensors. *Smart Materials and Stuctures* 9(3): 322.

Byrd, H. and Matthewman, S. (2014) Exergy and the City: The Technology and Sociology of Power (Failure). *Journal of Urban Technology* 21: 85-102.

Charlier, R.H. and de Meyer, C.P. (1992) An Environmental Purpose Artificial Island Off Belgium. *International Journal of Environmental Studies* 40: 249-265.

Caminos, H. and Cogliati-Bantz, V.P. (2014) *The Legal Regime of Straits: Contemporary Challenges and Solutions.* Cambridge: Cambridge University Press. 530 pages.

Cantrell, B.E. (2016) *Strategem for Responsive Technologies in Landscape Architecture.* New York: Routledge. 300 pages.

Capet, A. et al. (2016) Decline of the Black Sea oxygen inventory. *Biogeosciences* 13: 1287-1297/

Cathcart, R.B. (1983) Macroengineering Transformation of the Mediterranean Sea and Africa. *World Futures* 19: 111-121.

Cathcart, R.B. (18 August 1998) Land Art as global warming or cooling antidote. *Speculations in Science and Technology* 21: 65-72.

Cathcart, R.B. (April 2004) Architectural Ecology: A Tentative Sahara Restoration. *The International Journal of Environmental Studies* 61: 145-160.

Cathcart, R.B. (2013) Macro-Imagineering a 'Fifth Coast' USA: Salton Sea Megaport and Seawater Canal. Chapter 4, pages 127-156 ***IN*** White, M.R. (Ed.) *Seawater: Geochemistry, Composition and Environmental Impacts.* New York: NOVA Publishers. 176 pages.

Cathcart, R.B. (2015) Alexander the Great Sea Dam-Oceanarium Macroproject at Thessaloniki Bay, Greece: Potential Impacts of Massive Aquatecture. Chapter 1, pages 1-35 ***IN*** R.B. Cathcart, *World Macro-Imagineering Aquatecture Mega-projects.* West Conshokocken, PA: Infinity Publishing. 197 pages.

Cemen, I. and Yilmaz, Y. (2017) *Active Global Seismology: Neotectonics and Earthquake Potential of the Eastern Mediterranean Region.* New York: Wiley. 306 pages.

Chan, W-L and Motoi, T. (2003) Effects of stopping the Mediterranean Outflow on the southern polar region. *Polar Meteorology Glaciology* 17: 25-35.

Charlier, R.H. and Chaineux, M-C. P. (2009) The Healing Sea: A Sustainable Coastal Ocean Resource: Thalassotherapy. *Journal of Coastal Research* 25: 838-856.

Chen, J-B. et al. (November 2014) Anthropophile elements in river sediments: Overview from the Seine River. *Geochemistry, Geophysics, Geosystems* 11: 4526-4546.

Chianese, R.L. (September-October, 2013) Regeneration on *Tree Mountain. American Scientist* 101: 350-351.

Christianen, P.H. (2012) Dam Nation: Imaging and Imagining the "Middle East" in Herman Sorgel's Atlantropa. *International Journal of Islamic Architecture* 2: 325-345.

Churilova, T. et al. (4 April 2017) Light Absorption by Photoplankton in the Upper Mixed Layer of the Black Sea: Seasonality and Parametrization. *Frontiers in Marine Science* 4:90.

Connelly, J. (Winter 2012-2013) The Living Lighthouse. *Cabinet* Issue 48: 76-77.

Coulthard, T.J. et al. (September 2013) Were Rivers Flowing across the Sahara During the Last Interglacial? Implications for Human Migration through Africa. *PLOS ONE* 8: e74834.

Coward, F. (2016) Scaling up: Material culture as scaffold for the social brain. *Quaternary International* 405: 78-90.

Crespo-Blanc, M. et al. (30 June 2016) Clues for the Tortonian reconstruction of the Gibraltar Arc: Structural pattern, deformationism and block rotation. *Tectonophysis* 683: 308-324.

Debschitz, Uta von and Thilo von (2013) *FRITZ KAHN.* New York: Taschen. 392 pages.

Dellios, K. et al. (2015) Cyber Ocean: A roadmap to maritime cloud. *International Journal of Business Innovation and Research*9: 415-436.

De Preester, H. (2011) Technology and the Body: the (Im)Possibilities of Re-embodiment. *Foundations of Science* 16: 119-137.

Dick, J.T.A. et al. (2017) Functional responses can unify invasion ecology. *Biological Invasions* 19: 1667-1672.

Dobson, J.E. (April 2014) Aquaterra *Incognita*: Lost Land Beneath the Sea. *Geographical Review* 104: 123-138.

Donkers, J. et al. (1990) Strategies for Adaptation to Sea Level Rise. *Report, IPCC Coastal Zone Management Subgroup: Intergovernmental Panel on Climate Change.* Geneva, Switzerland. [GOTO: http://papers.risingsea.net/IPCC-1990-Strategies-for-Adaption-to-Sea-Level-Rise.html, 1990.]

Duarte, J.C. et al. (August 2013) Are subduction zones invading the Atlantic? Evidence from the southwest Iberia margin. *Geology* 41: 839-842.

Edgeworth, M. (April 2010) Beyond Human Proportions: Archaeology of the Mega and the Nano. *Archaeologies: Journal of the World Archaeology Congress* 6: 138-149.

El-Eraki, M.A. et al. (May 2016) Multi-seismotectonic models, present-day seismicity and seismic hazard assessment for the Suez Canal and its surrounding area, Egypt. *Bulletin of Engineering Geology and the Environment* 75: 503-517.

Elhacham, E. and Alpert, P. (2016) Impact of coastline-intensive anthropogenic activities on the atmosphere from moderate resolution imaging spectroradiometer (MODIS) data in Dubai (2001-2014).

Fend, P. (1999) *Ocean Earth: 1980-Today.* FRG: Oktagon Verlagsgesellschaft mbH. 272 pages.

Finkl, C.W. et al. (July 2012) A Review of Potential Tsunami Impacts to the Suez Canal. *Journal of Coastal Research* 28: 745-759.

Flaounas, E. et al. (February 2016) Cyclone contribution to the Mediterranean Sea water budget. *Climate Dynamics* 46: 913-927.

Foley, R.A. and Lahr, M.M. (2015) Lithic Landscapes: Early Human Impact from Stone Tool Production on the Central Saharan Environment. *PLoS ONE* 10: e0116482.

Forman, R.T.T. and Wu, J. (29 September 2016) Where to put the next billion people. *Nature* 537: 608-611.

Forootan, E. et al. (2017) Large-Scale Total Water Storage and Water Flux Changes over the Arid and Semiarid Parts of the Middle East from GRACE and Reanalysis Products. *Surveys in Geophysics* 38: 591-615.

Frainotti, D. et al. (2016) From dwindling ice to headwater lakes: could dams replace glaciers in the European Alps? *Environmental Research Letters* 11: 054022.

Fussl, W. (2010) The Deutsches Museum and its history. ***IN*** Heckl, W.M. (Ed.) *Technology in a Changing World: The Collections of the Deutsches Museum.* Translated by H. Casement and J. O'Mera. Munich, FRG: Deutsches Museum.

Galassi, G. and Spada, G. (December 2014) Sea-level rise in the Mediterranean Sea by 2050: Roles of terrestrial ice melt, steric effects and glacial isostatic adjustment. *Global and Planetary Change* 123, Part A: 55-66.

Galil, B. et al. (2017) The enlargement of the Suez Canal—Erythraean introductions and management challenges. *Management of Biological Invasions* 8: 1-12.

Galli, A. et al. (2017) Mediterranean countries' food consumption and sourcing patterns: An Ecological Footprint viewpoint. *Science of the Total Environment* 578: 383-391.

Galvin, J.F.P. (January 2010) When does it snow in the desert? *Royal Meteorological Society* 65(1): 10-13.

Gambolati, G. and Teatini, P. (2014) *Venice Shall Rise Again: Engineered Uplift of Venice Through Seawater Injection.* London: Elsevier. 83 pages.

Gandini, F. et al. (2016) Mapping human dispersals into the Horn of Africa from Arabian Ice Age refugia using mitogenomes. *Scientific Reports* 6: 25472.

Gerostathis, T.P. et al. (2016) 3D hydroelastic analysis of very large floating bodies over variable bathymetry regions. *Journal of Ocean Engineering Marine Energy* 2: 159-175.

Geymen, A. (March/April 2013) Impacts of Bosporus bridges on the Istanbul Metropolitan settlement areas. *Land Degradation & Development* 24: 156-169.

Giosan, L. et al. (2012) Early Anthropogenic Transformations of the Danube-Black Sea System. *Scientific Reports* 2: 582.

Goffredo, S. and Dubinsky, Z. (2014) *The Mediterranean Sea: Its History and Present Challenges.* The Netherlands: Springer. 678 pages.

Goldberg, S.L. et al. (15 October 2016) The timing of the Black Sea flood event: Insights from modeling of glacial isostatic adjustment. *Earth and Planetary Science Letters* 452: 178-184.

Goncalves, J. et al. (2013) Quantifying the modern recharge of the "fossil" Sahara aquifers. *Geophysical Research Letters* 40: 2673-2678.

Gouda, D.M. (2016) *Social Capital and Local Water Management in Egypt.* UK: Oxford University Press. 280 pages. [SEE also: Sarant, L. (27 April 2017) Space to grow. *Nature* 544: S14-S16.]

Gower, J. (September 2015) A sea surface height control dam at the Strait of Gibraltar. *Natural Hazards* 78: 2109-2120.

Grahn, H. et al. (15 September 2015) What's that smell? Hydrogen sulphide transport from Bardargunga to Scandinavia. *Journal of Volcanology and Geothermal Research* 303: 187-192.

Greene, C. et al. (December 2016) Marine Microalgae: climate, energy, and food security from the sea. *Oceanography* 29: 1-6.

Gupta, P. (November 2015) Futures, fakes and discourses of the gigantic and miniature in "The World" islands, Dubai. *Island Studies Journal* 10: 181-196.

Gurung, A.B. et al. (2016) Rethinking Pumped Storage Hydropower in the European Alps. *Mountain Research and Development* 36: 222-232.

Haff, P.K. (2014) Humans and technology in the Anthropocene: Six rules. *The Anthropocene Review* 1(2): 126-136.

Hafiez, R.A. (2011) Mapping of the Qattara Depression, Egypt, using SRTM Elevation Data for Possible Hydropower and Climate Change Macro-Projects. Pages 519-531 ***IN*** Badescu, V. and Cathcart, R.B. (Eds.) *Macro-Engineering Seawater in Unique Environments: Arid Lowlands and Water Bodies Rehabilitation.* The Netherlands: Springer. 790 pages.

Hafner, A. et al. (2017) Vertical mobility around the high-alpine Schnidejoch Pass: Indications of Neolithic and Bronze Age pastoralism in the Swiss Alps from paleoecological and archaeological sources. *Quaternary International.* DOI: 10.1016/j.quaint.2016.12.049.

Haklidir, M. et al. (2009) The possibilities for hydrogen production from H_2S and storage in the Black Sea. *International Journal of Nuclear Hydrogen Production and Applications* 2: 78-85.

Hall, J.K. (15 December 1996) Digital topography and bathymetry of the area of the Dead Sea Depression. *Tectonophysics* 266: 177-185.

Hamouda, A.Z. (2010) Worst scenarios of tsunami effects along the Mediterranean coast of Egypt. *Marine Geophysics Research* 31: 197-214.

Hansen, P. and Jonsson, S. (2014) *Eurafrica: The Untold History of European Integration and Colonialism.* New York: Bloomsbury Academic. 344 pages.

Harff, J. et al. (Eds.) (2016) *Geology and Archaeology: Submerged Landscapes of the Continental Shelf.* United Kingdom: Geological Society Special Publications. 294 Pages.

Hassan, El Sayed Mohamed (July 1961) Power from the sea. *Journal of the Power Division, Proceedings of the* American *Society of Civil Engineers* 87: 21-27.

Headley, J. M. (Spring 1997) The Sixteenth-Century Celebration of the Earth's Total Habitability: The Issue of the Fully Habitable World for Renaissance Europe. *Journal of World History* 8: 1-28.

Heartney, E. (October 2003) Mapping a Better World. *Art in America.* Pages 114-119.

Heefer, G. (2015) "A tract that is wholly sand": engineering military environments in Libya. *Endeavour* 40: 38-47.

Heffernan, O. (4 February 2016) The Mystery of the Expanding Tropics. *Nature* 530: 20-22.

Heness, E.A. et al. (2017) The shoreline distribution and degradation of tilapia carcasses, Salton Sea, California: Taphonomic implications. *Palaeogeography, Palaeoclimatology, Palaeoecology* 468: 276-286.

Hereher, M.E. (2015) Capacity assessment of the Qattara Depression: Egypt as a sink for the global sea level rise. *Geocarto International* 30: 123-131.

Hernandez-Molina, F.J. et al. (13 June 2014) Onset of Mediterranean outflow into the North Atlantic. *Science* 344: 1244-1250.

Homer, P.K. (April 2017) Genesis of a Commercial Space Suit from Dining Room Table to Revolutionary Space Company. *New Space.* DOI: 10.1089/space.2017.0003.

Hsu, P-C. et al. (2 September 2016) Radiative human body cooling by nanoporous polyethylene textile. *Science* 353: 1019-1023.

Jaffe, J.S. et al. (24 January 2017) A swarm of autonomous miniature underwater robot drifters for exploring submesoscale ocean dynamics. *Nature Communications* 8: 14189.

Italiano, F. et al. (2014) The Marsili Volcanic Seamount (Southern Tyrrhenian Sea): A Potential Offshore Geothermal Resource. *Energies* 7: 4068-4086.

Jiang, C. et al. (March 2017) Numerical Investigation of Tsunami-Like Solitary Wave Interaction with a Seawall. *Journal of Earthquake and Tsunami* 11: 1740006.

Johnson, R.G. (8 July 1997) Climate Control Requires a Dam at the Strait of Gibraltar. *EOS* 78(27): 277 and 280-282.

Jones, D.O.B. et al. (8 February 2017) Biological responses to disturbance from simulated deep-sea polymetallic nodule mining. *PLoS ONE* 12: 1-26.

Kahn, S.D. et al. (2014) Remote sensing and geophysical investigations of Moghra Lake in the Qattara Depression, Western Desert, Egypt. *Geomorphology* 207: 10-22.

Kander, A. and Malanima, P. (2015) *Power to the People: Energy in Europe over the Last Five Centuries.* Princeton: Princeton University Press. 472 pages.

Katz, A. and Starinsky, A. (2009) Geochemical History of the Dead Sea. *Aquatic Geochemistry* 15: 159-194.

Keller, S. and Jaeger, H.M. (May 2016) Aleatory architectures. *Granular Matter* 18: 1-11.

Kennett, D.J. and Kennett, J.P. (2006) Early State Formation in Southern Mesopotamia: Sea Levels, Shorelines, and Climate Change. *Journal of Island & Coastal Archaeology* 1: 67-99.

Kettani, M.A. and Gonsalves, L.M. (December 1972) Heliohydroelectric (HHE) power generation. *Solar Energy* 14: 29-39.

Khalil, B. et al. (2016) A review: dew water collection from radiative passive collectors to recent developments of active collectors. *Sustainable Water Resource Management* 2: 71-86.

Kirkby, M.J. (2016) Water in the critical zone: soil, water and life from profile to planet. *SOIL* 2: 631-645.

Kiro, Y. et al. (15 April 2017) Relationships between lake-level changes and water and salt budgets in the Dear Sea during extreme aridities in the Eastern Mediterranean. *Earth and Planetary Science Letters* 464: 211-226.

Klemm, O. et al. (2012) Fog as a Fresh-Water Resource: Overview and Perspectives. *AMBIO* 41: 221-234.

Koprulu, K. et al. (March 2016) Last Glacial-Holocene stratigraphic development at the Marmara Sea exit of the Bosphorus Strait, Turkey. *Marine Geophysical Research* 37: 21-35.

Kottegoda, N. et al. (2017) Urea-Hydroxyapatite Nanohybrids for Slow Release of Nitrogen. *ACS Nano*. DOI: 10.1021/acsnano.6b07781.

Kump, L.R. et al. (May 2005) Massive release of hydrogen sulfide to the surface ocean and atmosphere during intervals of oceanic anoxia. *Geology* 33: 397-400.

Laboratoire Bale (2017) *VENICE LESSONS: Industrial Nostaglia: Teaching and Research in Architecture.* FRG: Park Books. 216 pages.

Lagendijk, V. (2008) *Electrifying Europe: The Power of Europe in the Construction of Electricity Networks.* Amsterdam: Aksant. 246 pages.

Lambeck, K. et al. (December 2011) Sea level and shoreline reconstructions for the Red Sea: isostatic and tectonic considerations and implications for hominin migration out of Africa. *Quaternary Science Reviews* 30: 3542-3574.

Landers, R.B. (1966) Man's Place in the Dybosphere. New Jersey: Prentice-Hall. 261 pages. [GOTO: http://www.doaat.com/download.php?pid=34533550 for a free PDF of the book.]

Larkin, B. (2013) The Politics and Poetics of Infrastructure. *Annual Review of Anthropogology* 42: 327-343.

Larsson, M. and Kaiser, A. (2014) Invincible Cities for the Materiomic Age. Pages 53-97 ***IN*** Rassia, S.T. and Pardalos, P.M. (Eds.) *Cities for Smart Environmental and Energy Futures.* The Netherlands: Springer.

Leconte, J. et al. (12 December 2013) Increased insolation threshold for runaway greenhouse processes on Earth-like planets. *Nature* 504: 268-271.

Lefuvre, J-C. and Bouchard, V. (June 2002) From a civil engineering to an ecological engineering: An historical perspective from the Mont Saint Michael bay. *Ecological Engineering* 18: 593-606.

Lehmann, P.N. (2016) Infinite Power to Change the World: Hydroelectricity and Engineered Climate Change in the Atlantropa Project. *The American Historical Review* 121: 70-100.

Lelieveld, J. et al. (2016) Strongly increasing heat extremes in the Middle East and North Africa (MENA) in the 21st Century. *Climatic Change* 137: 245-260.

Liao, S.M. et al. (June 2012) Human Engineering and Climate Change. *Ethics, Policy and the Environment* 15: 206-221.

Lin, N. and Emanuel, K. (2016) Grey swan tropical cyclones. *Nature Climate Change* 6: 106-111.

Linnell, J.D.C. et al. (22 June 2016) Border Security Fencing and Wildlife: The End of the Transboundary Paradigm in Eurasia? *PLOS Biology* 14(6): e1002483.

Lokier, S.W. et al. (July 2015) Late Quaternary sea-level changes of the Persian Gulf. *Quaternary Research* 84: 69-81.

Lokman, K. (10 March 2017) Cyborg landscapes: Choreographing resilient interactions between infrastructure, ecology, and society. *Journal of Landscape Architecture* 12: 60-73.

Loring, P.A. (2007) The Most Resilient Show on Earth: The Circus as a Model for Viewing Identity, Change, and Chaos. *Ecology and Society* 12(1): 9.

Loughland, R.A. et al. (2012) Anthropogenic Induced Geomorphological Change Along the Western Arabian Gulf Coast. Chapter 8, page 211 ***IN*** Piacentini, T. (Ed.) *Studies on Environmental and Applied Geomorphology*. Croatia: INTECH. 294 pages.

Lowe, A. and Brotton, J. (2015) Revisioning the World: Mapping the Lithosphere. GOTO: www.factum-arte.com/download?f=lib/kcfinder/upload/files.

Machlis, G.E. and Hanson, T. (September 2008) Warfare Ecology. *BioScience* 58: 729-736.

Malkawi, A.I. and Tsur, Y. (2016) Reclaiming the Dead Sea: Alternatives for Action. Chapter 14, pages 205-225 ***IN*** Huttl, R.F., Bens, O., Bismuth, C. and Hoechstetter, S. (Eds.) (2016) *Society-Water-Technology: A Critical Appraisal of Major Water Engineering Projects.* The Netherlands: Springer.

Malkawi, A.I.H. et al (21-24 March 2010) The Red-to-Dead Sea conveyance system: bridging the water shortage and prospects for desalination. *Nuclear & Renewable Energy Conference (INREC), 2010 1st International.* 7 pages.

Marques, F.O. and Kaus, B.J.P. (January 2016) Speculations on the impact of catastrophic subduction initiation on the Earth System. *Journal of Geodynamics* 93: 1-16.

Marsh, P. (9 June 1977) Building islands for tomorrow's industry. *New Scientist* 74: 580-582.

Maselli, V. and Trincardi, F. (2013) Man made deltas. *Scientific Reports* 3: 1926.

Martin, A.G.L. (2014) *International Straits: Concept, Classification and Rules of Passage.* The Netherlands: Springer. 218 pages.

McLinden, C.A. et al. (2016) Space-based detection of missing sulfur dioxide sources of global air pollution. *Nature Geoscience* 9: 496-500.

Mensvoort, K.van and Grievink, H-J. (2011) *Next Nature: Nature Changes Along With Us.* Barcelona, Spain: ACTAR. 512 pages.

Meric, E. et al. (16 May 2016) Did the Red Sea-Mediterranean connection over the Dead Sea Fault end in the Late Pliocene? *Quaternary International* 401: 123-131.

Midilli, A. et al. (January 2007) A parametric investigation of hydrogen energy potential based on H_2S in Black Sea deep water. *International Journal of Hydrogen Energy* 32: 117-124.

Ming, T. et al. (2017) A moist air condensing device for sustainable energy production and water generation. *Energy Conversion and Management* 138: 638-650.

Mitchell, R.N. et al. (2012) Supercontinent cycles and the calculation of absolute palaeolongitude in deep time. *Nature* 482: 208-211.

Mitrovica, J.X. et al. (11 December 2015) Reconciling past changes in Earth's rotation with 20th century global sea-level rise: Resolving Munk's enigma. *Science Advances* 1: e1500679.

Moore, J.C. et al. (November 2016) Will China be the first to initiate climate engineering? *Earth's Future* 4: 1-8.

Morris, R.J. (1980) Floating debris in the Mediterranean. *Marine Pollution Bulletin* 11: 125.

Morton, M.Q. (2017) *Empires and Anarchie: A History of Oil in the Middle East.* London: Reaktion Books. 256 pages.

Mugnier, C.J. (October 2007) Grids & Datums: State of Bahrain. *Photogrammetric Engineering & Remote Sensing* 73: 1097.

Muller, K. (2000) The Birth and Death of Eurafrica. *International Journal of Francophone Studies* 3: 4-17.

Muller-Steinhagen, H. (2013) Concentrating solar thermal power. *Philosophical Transactions of the Royal Society A* 371: 20110433.

Naman, S.A. et al. (November 2008) Industrial extraction pilot plant for stripping H_2S gas from Black Sea water. *International Journal of Hydrogen Energy* 33: 6577-6585.

Naser, H.A. (February 2015) The role of environmental impact assessment in protecting coastal and marine environments in rapidly developing islands: The case of Bahrain, Arabian Gulf. *Ocean & Coastal Management* 104: 159-169.

Nash, D.J. et al. (15 December 2016) African hydroclimatic variability during the last 2000 years. *Quaternary Science Reviews* 154: 1-22.

Nerlich, B. and Jaspal, R. (2012) Metaphors We Die By? Geoengineering, Metaphors, and the Argument From Catastrophe. *Metaphor and Symbol* 27: 131-147.

Newell, R. and Canessa, R. (1 June 2017) Picturing a place by the sea: Geovisualizations as place-based tools for collaborative coastal management. *Ocean & Coastal Management* 141: 29-42.

Newman, W.S. and Fairbridge, R.W. (1986) The Management of sea-level rise. *Nature* 320: 319-321.

Ni, G. et al. (22 August 2016) Steam generation under one sun enabled by a floating structure with thermal concentration. *Nature Energy*. Article Number 16126. DOI: 10.1038/NERGY.2016.126.

Nomikou, P. et al. (2016) Post-eruptive flooding of Santorini caldera and implications for tsunami generation. *Nature Communications* 7: 13332.

Nordstrom, K.E. et al. (April 2008) The Marble Beaches of Tuscany. *The Geographical Review* 98: 280-300.

Orhon, D. (23 September 2014) Istanbul canal needs environmental study. *Nature* 513: 461.

O'Shea, S. (2017) *The Alps: A Human History from Hannibal to Heidi and Beyond.* New York: W.W. Norton.

Ostanciaux, E. et al. (January 2012) Present-day trends of vertical ground motion along the coast lines. *Earth-Science Reviews* 110: 74-92.

Othmani, A. et al. (May 2017) High-resolution numerical modelling of the barotropic tides of the Gulf of Gabes, eastern Mediterranean Sea (Tunisia). *Journal of African Earth Sciences* 129: 224-232.

Paez-Osuna, F. et al. (March 2017) Environmental status of the Gulf of California: A pollution review. *Earth-Science Reviews* 166: 181-205.

Papadopoulos, G.A. et al. (1 August 2014) Historical and pre-historical tsunamis in the Mediterranean and its connected seas: Geological signatures, generation mechanisms and coastal impacts. *Marine Geology* 354: 81-109.

Paris, P.J. et al. (16 December 2016) Where the continent ends. *Geophysical Research Letters* 43: 12208-12216.

Park, K-C. et al. (29 July 2013) Optimal Design of Permeable Fiber Network Structures for Fog Harvesting. *Langmuir* 29: 13269-13277.

Pasaric, M. et al. (2012) Historical tsunamis in the Adriatic Sea. *Natural Hazards* 61: 281-316.

Parolari, A.J. et al. (2016) Climate, not conflict, explains extreme Middle East dust storm. *Environmental Research Letters* 11: 114013.

Pearce, F. (6 April 2002) The sea of mirages. *New Scientist* 174: 44-45.

Piper, D.Z. (2016) Geochemistry of the Black Sea during the last 15 kyr: A protracted evolution of its hydrography and ecology. *Paleoceanography* 31: 1-21.

Pirni, A. (8 August 2016) Space and Anthropology of Limit: A Philosophical Perspective. *Frontiers in Astronomy and Space Sciences* 3: 22.

Piroddi, C. et al. (2017) Historical changes of the Mediterranean Sea ecosystem: modelling the role and impact of primary productivity and fisheries changes over time. *Scientific Reports* 7: 44491.

Pliego, J.M. (2005) The Gibraltar Strait tunnel: An overview of the study process. *Tunneling Underground Space Technology* 20: 558-569.

Povie, G. et al. (14 April 2017) Synthesis of a carbon nanobelt. *Science* 356: 172-175.

Powley, H.R. et al. (November 2016) Circulation and oxygen cycling in the Mediterranean Sea: Sensitivity to future climate change. *Journal of Geophysical Research: Oceans* 121: 8230-8247.

Previdi, M. et al. (2014) Climate system response to stratospheric ozone depletion and recovery. *Quarterly Journal of the Royal Meteorological Society* 140: 2401-2419.

Pugno, N.M. et al. (2011) Sea Art: The Mediterranean Sea Terrace Proposal. Volume 3, Pages 1425-1440 ***IN*** Brunn, S. D. (Ed.) *Engineering Earth: The Impacts of Megaengineering Projects*. The Netherlands: Springer. 2266 pages.

Qdais, H.A. (March 2008) Environmental impacts of the mega desalination projects: the Red-Dead Sea conveyor. *Desalination* 220: 16-23.

Quirke, S. (2009) Sehel and Suez: canal-cutting and periodization in ancient and modern history. Pages 223-230 ***IN*** Fitzenreiter, M. (Ed.) *Das Ereigns, Geschichtsschreibung zwischen Vorfall und Struktur.* London: Golden House Publications.

Ravazzoli, E. et al. (February 2017) The Effects of the Planned High-Speed Rail System on Travel Times and Spatial Development in the European Alps. *Mountain Research and Development* 37: 131-140.

Reba, M. et al. (June 2016) Spatializing 6,000 years of global urbanization from 3700 BC to AD 2000. *Scientific Data* 3: 160034.

Redmount, C.A. (April 1995) The Wadi Tumilat and the "Canal of the Pharaohs" *Journal of Near Eastern Studies* 54: 127-135.

Rickards, L. et al. (2015) Celebrating 80 years of the Permanent Service for Mean Sea Level (PSMSL). *Proceedings IAHS* 365: 1-5.

Rigby, M. et al. (17 April 2017) Role of atmospheric oxidation in recent methane growth. *Proceedings of the National Academy of Sciences*. DOI: 10.1073/pnas.1616426114.

Rivers, T.J. (2008) Technology's role in the confusion of needs and wants. *Technology in Society* 30: 104-109.

Robin, L. et al. (2014) Three galleries of the Anthropocene. *The Anthropocene Review* 1: 207-224.

Rodriguez, R.A. et al. (2017) Ecology: Science or philately? An interdisciplinary analysis of sustainability by exploring if it is possible to get more and more information by reducing collateral environmental damages. *Science of the Total Environment* 596: 43-52.

Romano, M. et al. (June 2016) The first geological map: an Italian legacy. *Italian Journal of Geoscience* 135: 1-6.

Roudaire, F.E. (May 1874) Une mer interieure en Algerie. *Revue des Deux Mondes*. Pages 323-360.

Roveri, M. et al. (2014) The Messinian Salinity Crisis: Past and Future of a great challenge for marine sciences. *Marine Geology* 352: 25-38.

Royakkers, L. and van Est, R. (November 2015) A Literature Review of New Robotics: Automation from Love to War. *International Journal of Social Robotics* 7: 549.

Rudy, A.P. (January 2005) Imperial contradictions: is the Valley a watershed, region, or cyborg? *Journal of Rural Studies* 21: 19-38.

Ruger, J. (2017) *Heligoland: Britain, Germany, and the Struggle for the North Sea.* New York: Oxford University Press. 370 pages.

Rugescu, R.D. et al. (2011) Macro-project on Electricity and Freshwater Supply for Djibouti, Lac Assal Area. *Proceedings of the 8th International Conference on Structural Dynamics, EURODYN 2011, Leuven, Belgium, 4-6 July 2011. Paper #351. Pages 3582-3588.*

Rumpf, C.M. et al. (2017) Population Vulnerability Models for Asteroid Risk Assessment. *Meteoritics & Planetary Science.* DOI: 10.1111/maps.12861.

Rumpf, C.M. et al. (2017) Asteroid impact effects and their immediate hazards for human populations. *Geophysical Research Letters* 44: 1-8. DOI: 10.1002/2017GL073191.

Salter, S.H. et al. (2014) Engineering Ideas for Brighter Clouds. *Issues in Environmental Science and Technology* 38: 131-161.

Sanderson, E.W. (2009) *Mannahatta: A Natural History of New York City.* New York: Abrams. 352 pages.

Sannino, G. et al. (April 2017) A high-resolution modelling study of the Turkish Straits System. *Ocean Dynamics* 67: 397-432.

Sarant, L. (1 September 2016) An end to oil dependency. *Nature* 537: 56-57.

Scanlon, B.R. and Smakhtin, V. (2016) Focus on water storage for managing climate extremes and change. *Environmental Research Letters* 11: 120208.

Schneck, R. et al. (2010) Climate modelling sensitivity experiments for the Messinian Salinity Crisis. *Palaeogeography, Palaeoclimatology, Palaeoecology* 286: 149-163.

Schuiling, R.D. et al. (January-June 2005) The Hormuz Strait Dam Macroproject—21st Century Electricity Development Infrastructure Node (EDIN)? *Marine Georesources and Geotechnology* 23: 25-37.

Schuiling, R.D. et al. (2007) Power from closing the Red Sea: economic and ecological costs and benefits following the isolation of the Red Sea. *International Journal of Global Environmental Issues* 7: 341-361.

Schweizer, J. et al. (1 September 2016) Investigating the potential and feasibility for an offshore wind farm in the Northern Adriatic Sea. *Applied Energy* 177: 449-463.

Schwendtner, B. et al. (2013) Risk evolution: how can changes in the built environment influence the potential loss of natural hazards? *Natural Hazards and Earth System Science* 13: 2195-2207.

Sepic, J. et al. (2012) Northern Adriatic meteorological tsunamis: Observations link to the atmosphere, and predictability. *Journal of Geophysical Research* 117: 1-18.

Sheppard, C. et al. (2010) The Gulf: a young sea in decline. *Marine Pollution Bulletin* 60: 13-38.

Shi, N.N. et al. (17 July 2015) Keeping cool: Enhanced optical reflection and heat dissipation in silver ants. *Science* 349: 298-301.

Shim, J-S and Yoon, J-J. (2017) Hydraulic Experiments on Sinkable Floating Storm Surge Barrier under Various Wave Actions. *Journal of Coastal Research* 79: 184-188.

Showers, K.B. (2014) Europe's long history of extracting African renewable energy: Contexts for African scientists, technologists, innovators and policy-makers. *African Journal of Science, Technology, Innovation and Development* 6: 301-313.

Smith, R. et al. (2007) Sensitivity of hypersaline Arabian Gulf to seawater desalination plants. *Applied Mathematical Modelling* 31: 2347-2354.

Soliman, G.F. et al. (2001) Vulnerability Assessment of the Coastal Area in the Southern Part of the Suez Canal due to Global Warming and Accelerated Sea Level Rise. *Journal of KAU: Marine Science* 12: 49-82.

Spada, G. et al. (1999) Chandler wobble excitation by catastrophic flooding of the Black Sea. *Annali Di Geofisica* 42: 749-754.

Stanley, J-D. (2017) Increased Land Subsidence and Sea-Level Rise are Submerging Egypt's Nile Delta Coastal Margin. *GSA Today* 27: 1-8. DOI: 10.1130/GSATG312A.1.

Starr, V.P. and Anati, D.A. (April 1971) The Earth's gaseous hydrosphere as a natural resource. *Hydrology Research* 2: 65-78.

Starr, V.P. et al. (20 September 1974) Effectiveness of controlled convection in producing precipitation. *Journal of Geophysical Research* 79: 4047-4052.

Steinberg, T. (2014) *GOTHAM UNBOUND: The Ecological History of Greater New York.* New York: Simon & Schuster. 517 pages.

Stern, E. and Gradus, Y. (1983) The Negev continental bridge: a chain in an intermodal transport system. *Geoforum* 14: 461-469.

Stigebrandt, A. (2016) Restoration of the Baltic Proper by decadal oxygenation of the deepwater. *Ocean Science Discussions* 1-13. DOI: 10.5194/os-2016-17, 2016.

Sun, X. et al. (2012) The current state of offshore wind energy technology development. *Energy* 41: 298-312.

Tabbert, Thomas T. (2005) *Fused with the Absolute Reality Machine: Oswald's Wiene's "The Improvement of Central Europe" novel.* Hamburg, FRG: Artislife Press eK. 190 pages.

Tamburrino, S. et al. (October 2015) The proximal marine record of the Marsili Seamount in the last 7 ka (Southern Tyrranean Sea, Italy): Implications for the active processes in the Tyrrhanean Sea back-arc. *Global and Planetary Change* 133: 2-16.

Tarolli, P. and Sofia, G. (2016) Human topographic signatures and derived geomorphic processes across landscapes. *Geomorphology* 255: 140-161.

Taye, M. et al. (November 2016) The Grand Ethiopian Renaissance Dam: Source of Cooperation or Contention? *ASCE Journal of Water Resources Planning Management* 142.

Teller, J.T. et al. (2000) Calcareous dunes of the United Arab Emirates and Noah's Flood: the postglacial reflooding of the Persian (Arabian) Gulf. *Quaternary International* 68-71: 297-308.

Tepper, J.H. (1999) Connecting Geology, History, and the Classics Through a Course in Geomythology. *Journal of Geoscience Education* 47: 221-226.

Tesfamichael, D. and Pauly, D. (Eds) (2016) *The Red Sea Ecosystem and Fisheries.* The Netherlands: Springer. 77 pages.

Tolkova, E. (2016) Tsunami penetration in tidal rivers, with observations of the Chile 2015 tsunami in rivers in Japan. *Journal of Pure and Applied Geophysics* 173: 389-409.

Trezzi, G. et al. (1 March 2017) Assessing the role of submarine groundwater discharge as a source of Sr to the Mediterranean Sea. *Geochimica et Cosmochimica Acta* 200: 42-54.

Trigger, B.G. (1990) Monumental architecture: a thermodynamic explanation of symbolic behavior. *World Archaeology* 22: 119-132.

Trincardi, F. et al. (December 2016) The 1966 Flooding of Venice. *Oceanography* 29: 178-186.

Turner, F. (1998) Terraforming and the Coming Charm Industries. *Advances in Space Research* 22: 433-439.

Vernadsky, V. (1944) A Few Words About the Noosphere. *The Progress of Modern Biology* 18: 113-120.

Vitaliano, D.B (1973) *Legends of the Earth: Their geological origins.* Bloomington: Indiana University Press.

Weaver, A. (2005) Spaces of Containment and Revenue Capture: "Super-Sized" Cruise Ships as Mobile Tourism Enclaves. *Tourism Geographies: An International Journal of Tourism Space, Place and Environment* 7: 165-184.

Wegwerth, A. et al. (1 March 2016) Northern hemisphere climate control on the environmental dynamics in the glacial Black Sea "Lake". *Quaternary Science Reviews* 135: 41-53.

Weisman, A. (2007) *The World Without Us*. New York: Thomas Dunne Books. 324 pages.

Werner, F. and Lange, K. (1975) A bathymetric survey of the sill area between the Red Sea and the Gulf of Aden. *Geologisches Jahrbuch D* 13: 25-30.

Werner, L. (September-October 2007) A Cypress in the Sahara. *Saudi Aramco World* 58: 32-39.

Wiener, O. (Summer 2001) Remarks on Some Tendencies of the "Vienna Group". *OCTOBER* 97: 120-130. [A 23-page brochure, *Der naturliche und der kunstliche Mensch. Oswald Wiener: appexdix a: der bio-adapter*, focused on Wiener's Bio-Adapter was produced by S. Holzheimer in 2002 and published by GRIN Verlag as an e-book.]

Worzyk, T. (2009) *Submarine Power Cables: Design, Installation, Repair, Environmental Aspects*. The Netherlands: Springer. 296 pages.

Wright, D.K. (2017) Humans as Agents in the Termination of the African Humid Period. *Frontiers in Earth Science* 5:4: 1-14.

Wu, J. et al. (1 May 2017) North-African paleodrainage discharges to the central Mediterranean during the last 18,000 years: A proxy characterization. *Quaternary Science Reviews* 163: 95-113.

Xue, Y. and Shukla, J. (December 1996) The Influence of Land Surface Properties on Sahel Climate. Part II: Afforestation. *Journal of Climate* 9: 3260-3275.

Yagil, Y. et al. (January 1985) Near Drowning in the Dead Sea: Electrolyte Imbalances and Therapeutic Implications. *Archive of Internal Medicine* 145: 50-53.

Yang, D. et al. (2016) New road for telecoupling global prosperity and ecological sustainability. *Ecosystem Health and Sustainability* 2: e01242.

Zalasiewicz, J. (2008) *The Earth After Us: What Legacy will Humans Leave in the Rocks?* New York: Oxford University Press. 251 pages.

Zerbini, S. et al. (April 2017) Sea-level change in the Northern Mediterranean Sea from long-period tide gauge time series. *Earth-Science Reviews* 167: 72-87.

Zevenbergen, C. (2017) *Amphibious Building Design and Construction.* Florida: CRC Press. 350 pages.

Zhang, G.J. et al. (2013) Energy consumption and the unexplained winter warming over northern Asia and North America. *Nature Climate Change* 3: 466-470.

Zhang, Z. et al. (18 September 2014) Aridification of the Sahara desert caused by Tethys Sea shrinkage during the Late Miocene. *Nature* 513: 401-404.

Zhiltskov, S.S. et al. (Eds.) (2016) *Oil and Gas Pipelines in the Black-Caspian Seas Region*. The Netherlands: Springer. 284 pages.

Made in the USA
Monee, IL
04 November 2026

67386407R00115